GRP IN STRUCTURAL ENGINEERING

GRP IN STRUCTURAL ENGINEERING

the late M. HOLMES

and

D. J. JUST

*Department of Civil Engineering,
The University of Aston in Birmingham, UK*

APPLIED SCIENCE PUBLISHERS
LONDON and NEW YORK

APPLIED SCIENCE PUBLISHERS LTD
Ripple Road, Barking, Essex, England

Sole Distributor in the USA and Canada
ELSEVIER SCIENCE PUBLISHING CO., INC.
52 Vanderbilt Avenue, New York, NY 10017, USA

British Library Cataloguing in Publication Data

Holmes, M.
 GRP in structural engineering.
 1. Glass reinforced plastics
 I. Title. II. Just, D. J.
 624.1'89'23   TA455.P55

ISBN 0-85334-232-6

WITH 126 ILLUSTRATIONS AND 16 TABLES

© APPLIED SCIENCE PUBLISHERS LTD 1983

All rights reserved. No part of this publication may be reproduced, stored in a retrieval system, or transmitted in any form or by any means, electronic, mechanical, photocopying, recording, or otherwise, without the prior written permission of the copyright owner, Applied Science Publishers Ltd, Ripple Road, Barking, Essex, England

Photoset in Malta by Interprint Ltd
Printed in Great Britain by Galliard (Printers) Ltd, Great Yarmouth

Preface

The advances made by organic chemists during recent decades have led to the emergence of the new class of man-made materials commonly known as plastics, but more precisely termed super-polymers.

The possibilities of using these plastic materials in engineering situations are now being extensively examined, and in the field of structural engineering such development is taking place mainly in their use as glass fibre-reinforced plastics, the plastic material most widely used being polyester resin. Thus the abbreviation 'GRP', although usually thought of as meaning 'glass-reinforced plastics', should strictly be interpreted as 'glass-reinforced polyesters'.

The different forms, both in kind and degree, that such fibre-reinforced materials may take when compared with the more traditional materials have necessitated the development of analytical and design procedures which have not been generally required in treatments involving the more standard structural media.

Glass-reinforced plastics have now been used in the structural engineering field in a variety of applications, in particular for lightly loaded roofs, for pipes, and even for chimneys, and their range of application is growing as their nature and behaviour are becoming better appreciated.

In the absence, at present, of any Codes of Practice regarding the design of GRP structures, design in the medium is very highly dependent upon the results of research and on the experience gained from previous designs. Thus it is the main intention of this book to present the various relevant structural properties of glass fibre-reinforced plastics in as quantitative a manner as possible, at the same time aiming to acquaint the reader with the basic nature of GRP, to develop the fundamental theoretical background, and to describe the design process in the material.

To attempt to accomplish these ends, the book is divided into three parts which, although linked, may be read independently, the contributions from the first author (M. H.) being mainly to the experimental and design aspects found in Part C, while the second author's (D. J. J.) contributions are predominantly to the theoretical developments of Part B. Part A serves as an introduction to GRP from a historical and manufacturing viewpoint.

The book may thus be read from differing standpoints. Either it may be primarily considered as containing information regarding the properties of the material and the design of GRP structures, many of the results presented being the result of recent research carried out in the Department of Civil Engineering at the University of Aston, or it may be used as an introduction to orthotropic material and laminate behaviour, the applications of which can include composite materials other than GRP.

No knowledge beyond that acquired during the first and second years of a civil engineering degree course is required, although an acquaintance with basic matrix operations may be found beneficial to an understanding of some of the theoretical treatment.

It is greatly to be regretted that the death of Professor Malcolm Holmes in 1982 prevented him from seeing this book completed, although at that time Part C, with which he had been predominantly concerned, was almost complete. However, I know that he would wish to join me in acknowledging Messrs Pilkington Bros for their generous supply of glass fibre reinforcement and for their guidance regarding its fabrication, Miss Joan P. Stanley for her excellent typing of a sometimes difficult manuscript, and Mrs Sylvia Lancaster for her painstaking preparation of the illustrations. Finally gratitude is expressed to the authors' wives, Ruby and Patricia, for their patience and encouragement.

D. J. JUST
University of Aston in Birmingham

Contents

Preface v

Principal Notation xi

PART A: THE NATURE AND PRODUCTION OF GRP

Chapter 1 **Historical Background** 3
    1.1 The Requirements of Structural Materials 3
    1.2 Influence of the Nature of Materials on Structural Form ... 5
    1.3 The Nature of Structural Materials 6
        1.3.1 Homogeneous materials 6
        1.3.2 Composite materials 8
    References 13

Chapter 2 **The Manufacture and Properties of the Components of Glass-reinforced Polyesters** 14
    2.1 The Polyester Matrix 14
        2.1.1 The basic chemistry of polyester resins 14
        2.1.2 The production of unsaturated polyester resins . 18
        2.1.3 The properties of cured unsaturated polyester resins 20
    2.2 The Glass Fibre Reinforcement 23
        2.2.1 The structure and types of glass 24
        2.2.2 The production of glass fibres 26
        2.2.3 The properties of glass fibres 27
        2.2.4 The forms of glass fibre reinforcement 28
    References 30

Chapter 3 **The Manufacture and Characteristics of Glass-reinforced Polyesters** 32
    3.1 The Methods of Manufacture 32

	3.1.1 Open mould processes	32
	3.1.2 Closed mould processes	38
3.2	Material Characteristics	41
References		43

PART B: THE STRUCTURAL PROPERTIES OF GRP

Chapter 4 The Theoretical Macromechanical Properties of Composite Laminae 47

4.1	Introduction	47
4.2	The Assumptions and Idealisations Made in the Theoretical Treatment of Composite Laminae	48
4.3	The Stress–Strain Relationships for Composite Laminae	49
	4.3.1 Isotropic lamina	49
	4.3.2 Orthotropic lamina — principal orientation	61
	4.3.3 Orthotropic lamina — arbitrary orientation	71
4.4	Strength Characteristics of Composite Laminae	87
	4.4.1 The basic concepts in the determination of strength characteristics	88
	4.4.2 Strength hypotheses for isotropic laminae	89
	4.4.3 Strength hypotheses for orthotropic laminae	98
References		104

Chapter 5 The Theoretical Micromechanical Analysis of Composite Laminae 105

5.1	Introduction	105
5.2	Assumptions and Limitations in Micromechanical Analyses	106
5.3	The Stiffness Characteristics of Glass-reinforced Laminae	107
	5.3.1 Stress–strain relationship in a continuous unidirectional fibre lamina	107
	5.3.2 Stress–strain relationships in a discontinuous fibre lamina	120
5.4	The Strength Characteristics of Glass-reinforced Laminae	124
	5.4.1 Strengths of a continuous unidirectional fibre lamina	125
	5.4.2 Strengths of a discontinuous fibre lamina	130
References		132

Chapter 6 The Behaviour of Glass Fibre-reinforced Laminates ... **134**

6.1	Introduction	134
6.2	Stiffness Characteristics of Laminated Composites	135
	6.2.1 Behaviour of laminated beams	136

		6.2.2 Behaviour of laminated plates	143
	6.3	Strength Characteristics of Laminated Composites	164
		6.3.1 Strength analysis and failure criteria	166
	6.4	The Effect of Interlaminar Stresses	168
	References		169

Chapter 7 The Theoretical and Measured Properties of Glass-Reinforced Composites ... **171**

- 7.1 Introduction ... 171
- 7.2 Continuously Reinforced Laminates ... 171
  - 7.2.1 Unidirectionally continuously reinforced laminates ... 174
  - 7.2.2 Multidirectionally continuously reinforced laminates ... 182
- 7.3 Discontinuously Reinforced Laminates ... 209
  - 7.3.1 Stiffness and strength properties of discontinuously reinforced laminates ... 209
- References ... 212

Chapter 8 Time and Temperature Dependent Characteristics of Glass-reinforced Plastics ... **213**

- 8.1 Introduction ... 213
- 8.2 Basics of Linear Viscoelasticity ... 214
  - 8.2.1 Voigt model ... 215
  - 8.2.2 Maxwell model ... 218
  - 8.2.3 Combination of Voigt and Maxwell models ... 220
- 8.3 Dependence of the Stiffness of GRP Composites upon Time and Temperature ... 221
- 8.4 Dependence of the Strength of GRP Composites upon Time and Temperature ... 226
- References ... 228

PART C: DESIGN IN GRP

Chapter 9 Properties of GRP Relevant to Structural Design ... **233**

- 9.1 Introduction ... 233
- 9.2 Short Term Tensile and Compressive Strength and Stiffness ... 234
- 9.3 Short Term Flexural Strength and Stiffness ... 237
- 9.4 Short Term Shearing Strength and Stiffness ... 238
- 9.5 Long Term Strength and Stiffness Properties of GRP ... 238
- 9.6 Temperature Effects on GRP ... 241
- 9.7 Performance of GRP Structures in a Fire ... 241
- 9.8 Structural Joints or Connections ... 243
  - 9.8.1 Adhesive joints ... 243
  - 9.8.2 Mechanical joints ... 244

	9.8.3 Combination joints	247
	9.9 Structural Properties of GRP Relative to Other Structural Materials	248
	9.10 Transformed Sections	253
	References	257

Chapter 10 **Design of GRP Box-beams** **258**

- 10.1 Introduction 258
- 10.2 Loading, Span and Cross-sectional Shape 258
- 10.3 Selection of GRP Material 260
- 10.4 Beam Manufacture 261
- 10.5 Transformed Section and Critical Buckling Stresses for Design 263
- 10.6 Beam Stresses 268
- 10.7 Experimental Behaviour of the Beams 271
- 10.8 Effect of GRP Material Properties on Beam Performance 272
  - 10.8.1 Modulus of elasticity 272
  - 10.8.2 Compressive strength 273
  - 10.8.3 I value of transformed section 273
  - 10.8.4 Prevention of compression buckling failure..... 273
- 10.9 Behaviour of Beams under Long Term Loading 274
- References 280

Chapter 11 **Design of a Stressed Skin Roof Structure** **281**

- 11.1 Introduction 281
- 11.2 Loading and Material Properties 283
- 11.3 Preliminary Design 286
- 11.4 Computer Analysis 288
- References 291

Index 293

Principal Notation

A	Cross-sectional area	V_f	Fibre volume fraction
b	Breadth	w	Displacement along z-axis; load per unit run
d	Depth; diameter of fibre		
E	Young's modulus (modulus of elasticity)	W	Total load
		x, y, z	Orthogonal coordinates
g	Gravitational acceleration	X	Normal strength in 1-direction
G	Modulus of rigidity		
I	Second moment of area	Y	Normal strength in 2-direction
J	Torsional constant		
L	Length or span	Z	Normal strength in 3-direction; section modulus
l	Length of discontinuous fibre		
M	Bending moment; bending moment per unit width	1, 2, 3	Principal material directions in orthotropic material
N	Axial force per unit width		
P	Axial force	γ	Shearing strain
$[Q]$	Material matrix	δ	Deflection
S	Shearing strength relative to 1,2 directions	ε	Normal strain
		$[\varepsilon_0]$	Vector of axial strains
$[S]$	Compliance matrix	η	Coefficient of linear viscosity
t	Thickness; time		
T	Temperature	θ	Angle of rotation
u	Displacement along x-axis	$[\chi]$	Vector of flexural strains
U	Strain energy per unit volume	ν	Poisson's ratio
		ρ	Density
v	Displacement along y-axis	σ	Normal stress
V	Shearing force	τ	Shearing stress

Part A

THE NATURE AND PRODUCTION OF GRP

CHAPTER 1

Historical Background

1.1 THE REQUIREMENTS OF STRUCTURAL MATERIALS

The selection of the most suitable materials to adopt in the construction of any engineering project is so fundamental to the design process that it is not in the least surprising that a vast number of materials, both natural and artificial, has been experimented with, developed, and used in technological works of all kinds, and that as the requirements of engineering products have become greater and more diverse, so this spectrum has been increased by the development of new materials able satisfactorily to meet these further demands.

What then are the essential requirements of structural materials? Above all others, two fundamental properties are of paramount importance in the vast majority of applications. The first of these is that the material should have adequate *strength*. By this is meant that the material must be able to resist the various forces applied to it without some form of failure occurring. This strength is a function of the magnitude of the internal forces of attraction between the fundamental particles of the material, and the importance of strength is readily appreciated when one considers the magnitude of the forces that many of our present-day structures are required to withstand.

The second requirement is that the material should possess adequate *stiffness*. In the majority of engineering structures it is essential that shape is retained under the various combinations of loading to which they are to be subjected. This implies that the materials when subjected to force must have a high resistance to deformation, or in other words the deformations caused by the applied forces should be small. This relationship between the forces and the deformations caused is a measure of the stiffness of the material and hence, in general, materials of high stiffness are preferred in structural applications. It is also desirable that the force–deformation relationship may be considered the same under either loading or unloading conditions so that upon removal of load the deformations so caused should disappear. When this occurs the structure is said to be behaving *elastically*. Also calculations are eased and a ready understanding of the mode of structural behaviour obtained if the ratio between the loading and the displacement caused is constant. If this property applies to a material, in addition to showing elastic behaviour, the material is said to be *linear elastic*.

Although these two characteristics of strength and stiffness are the fundamental properties that a structural material must generally possess, a number of other requirements should also be given some consideration.

Firstly, it follows from the previous discussion on stiffness and elasticity, that the material should, as far as possible, cease to deform once the loading has been applied. If deformation continues to increase for some time after the load has been applied the material is said to be undergoing *creep* which, if significant, could lead to unsightly and perhaps dangerous deformations in the structure as well as to a possible reduction in the strength of the material. Creep occurs in a number of materials to a greater or lesser extent and becomes increasingly significant at high stresses or temperatures.

Secondly, the material should possess a specific gravity, or density, adequate for the design functions to be fulfilled. In many cases, the loading to which a structure is subjected is required to be minimised and thus quite often the property of low density in a material is advantageous. In other instances materials of high density are beneficial to enable the self weight of the structure alone to prevent or minimise undesirable reversal effects possible when variable 'live' loads become operative. In addition, when the number of the stress reversals to which the structure is subjected is large, the adequacy of the fatigue properties and perhaps of the impact resistance of the material should be examined.

Supplementary to these basic properties the material should be able to 'fail safely'. Although this appears to be a contradiction, ideally materials should be used that give warning of approaching failure, for instance by the onset of large deflections rather than by a sudden collapse. By the same token, in case of fire, materials prone to sudden combustion or to the production of noxious fumes should be avoided. Also the material should exhibit resistance to any other forms of deterioration to which it may be subjected.

In summarising, therefore, over the range of temperatures to which they are to be exposed, structural materials should generally possess both high strength and stiffness, at the same time being of suitable density, fairly resistant to creep and also able to give sufficient warning of impending failure.

1.2 INFLUENCE OF THE NATURE OF MATERIALS ON STRUCTURAL FORM

The physical form that a structure may take is fundamentally linked to the properties of the materials from which it is constructed, as a few simple examples will demonstrate. For instance, for long span beam structures sustaining large loads and where flexure is predominant, materials with high tensile and compressive strength properties are required, as well as with adequate shearing strength. In addition, to minimise deflections, the materials should have a high stiffness and, in order to reduce the self weight loading, low density materials would be advantageous. Materials with excessive creep properties should naturally be avoided. If, however, a material which, although possessing a high compressive strength, exhibits a low tensile resistance is considered, then the structure must take a form such that the tensile stresses in it will be low and the loads will be carried mainly by compression. The classical example of this class of structure is the arch, which can be so successfully constructed in brickwork, a medium offering little tensile resistance especially at the brick boundaries.

On the other hand, materials of low stiffness and with perhaps less stringent creep characteristics may be used in comparatively lightly loaded constructions and in structures where flexure is minimal. The shell roof is a typical example of such an application, the forces, as in the arch, being predominantly carried by compression and the loading consisting of the self weight of the roof and the comparatively light 'live' load due to snow and wind. Again, in order to reduce the magnitude of the loading, low density materials are required. If the material has a comparatively high tensile strength, then it may be advantageous to construct an inverted shell, as then compression and hence any inherent buckling problems are eliminated.

From these elementary considerations it is apparent that the nature of the material may have a profound effect on the form which the structure may take.

1.3 THE NATURE OF STRUCTURAL MATERIALS

What then are the materials available for structural use? On reflection it is noticed that very few possessing the necessary requirements occur naturally, only perhaps timber and stone or rock being directly useable, and the latter certainly in a limited fashion. Indeed, timber itself must be formed from its natural state in order to be structurally viable.

Thus it is observed that virtually every material having structural capabilities must be, to a greater or lesser extent, an artificial product, and that such materials have been and are continually being produced to meet the requirements of engineering developments.

A closer examination of these materials reveals that they can be considered in two broad groups. The first comprises what may be termed 'homogeneous' materials, that is materials whose compositions may be considered macroscopically uniform throughout; the second, generally newer class, consists of materials with two or more distinct homogeneous components, these being designated 'composite' materials.

1.3.1 Homogeneous Materials

There can be little doubt that the most widely used materials of this class are those based on iron. Iron is a most abundant element, forming about 5% of the earth's crust in the form of its compounds, and although it has been used for perhaps 3000 years, its exploitation and importance as a major structural material developed with the Industrial Revolution at the beginning of the nineteenth century. The requirements of the new machinery, particularly the steam engine, made imperative the cheap production of a material capable of withstanding the higher stresses and temperatures that were being imposed, and thus the production of iron and new engineering developments progressed side by side.

The first materials to be obtained from the pig iron produced by the blast furnace were cast iron and wrought iron. Cast iron, the cheaper to produce, was used extensively during the nineteenth century and even earlier, as the world's first metal bridge at Ironbridge in Shropshire suggests. The major defect of cast iron as a structural material is its low tensile strength and its brittle nature, and thus designs in cast iron should be such that the various structural elements are subjected to compression with fairly large factors of safety so that tensile stresses set up due to any buckling tendencies will be very small. The great advantage of cast iron is, as its name implies, that it can be cast in moulds to any desired shape, although of course high temperatures are required for the process and the production of the mould itself may demand highly skilled techniques.

Wrought iron, on the other hand, possesses much greater ductility, and hence the tensile strength is more effectively utilised. Unlike cast iron it is not easily moulded to shape and is used mainly in the form of rolled sections.

Although cast and wrought iron are by no means obsolete as engineering materials, and although they both have excellent resistance to corrosion, the two remarkable discoveries of Bessemer and Siemens rapidly led to the decline of iron as a structural medium, and to its replacement by one of the now most common and all-embracing materials, steel. The Bessemer and Siemens processes rapidly remove, by oxidation, the impurities from the molten iron, and by the addition of controlled amounts of carbon and other alloying elements to the treated metal, a ductile material of exceptionally high tensile strength is created. Like wrought iron, structural steel products are predominantly of rolled sections or plate, and indeed the new material was originally a substitute for the wrought iron used for railways and ship-building. The large-scale production of steels changed the face of all engineering developments. The elimination of brittleness from ferrous materials and the increase in ductility led to safer products, and the tremendous increase in tensile strength enabled structures incorporating both tensile and flexural components to be utilised. Steel possesses virtually all the requirements of a structural engineering material. It exhibits high tensile and compressive strength, has a high stiffness, and has a large range of linear elastic behaviour at normal temperatures with negligible creep. In addition, when strained above its elastic limit it does not suddenly fail but undergoes large deformations without an increase in stress. This 'plastic' property implies that a steel structure will not fail suddenly, as would one of cast iron, but will exhibit large deformations before actual collapse occurs, thus giving adequate warning of impending failure.

The adoption of steel in structural engineering produced enormous changes in design and architecture, and allowed for the construction of clear spans that could never have been considered previously. These achievements are witnessed today in the enormous suspension bridges and roof spans that have been produced in recent years. The 'plastic' property of steel has also over the last four decades initiated new concepts and techniques in structural analysis and design which make use of the large deformations that occur suddenly when the elastic limit is exceeded and the material 'yields'.

Although the use of steel is predominant in structural engineering, its tendency to corrode, unless adequately protected or unless the expensively-produced stainless steel is used, and its relatively high density are disadvantages in certain applications. For instance, this century has witnessed the development of the aircraft, a structure which should be not only capable of withstanding large forces but also as light as possible. These requirements have meant that materials less dense than steel but which still exhibit high strength and stiffness are required. In the late nineteenth century such a material became commercially viable, and has been used extensively since. This was aluminium, which like iron can be alloyed with other elements to produce high strength alloys. Aluminium alloys have a density and a stiffness about one-third that of steel whilst possessing fairly high tensile and compressive characteristics, and unlike common steel they have relatively good non-corrosive properties.

Although other metals, such as titanium and copper, have some structural applications, it is iron, steel, and to a lesser extent aluminium alloys that have been the homogeneous materials that have dominated engineering construction during the past two centuries. The basic structural requirements are nearly all satisfied by these materials and therefore their rapid decline in use is inconceivable.

However, although these materials have such good qualities, they lack one important property, namely the ability to be easily and economically formed into any shape, and it is this lack that has led this century to the development of an even more artificial class of materials known collectively as 'composites'.

1.3.2 Composite Materials

Although the materials already considered were termed 'homogeneous', in fact, as has been described, they usually consist of more than one constituent due to an alloying or additive process and hence are strictly composite materials. However, the matrices of which they are composed are randomly composite on a microscopic scale and hence they can be considered homogeneous as far as the determination of their mechanical properties is concerned.

True composite materials are those which are composed of more than one such homogeneous material, so that the properties of the composite are different from those of the individual constituents, and are in the great majority of cases man-made.

These artificially produced materials have been developed during the past hundred years in response to the limiting factors inherent in the homogeneous materials, and to the architectural, analytical, and design innovations that have developed.

Although such materials are predominantly artificial in nature, there is one composite material which has been used in construction before any other material so far considered, namely timber. Wood is in fact composed of cellulose tubes bonded together by another material, lignin, and the difference between this and the structure of steel which is considered to be homogeneous is quite distinct. Hence the concept of composite materials is not primarily the product of man's ingenuity, but is of natural origin.

The concept of producing a composite construction medium artificially dates back to the Roman, or even Egyptian, periods when mortars and cements made predominantly from lime, sand and clay were used to bond together pieces of stone or gravel. These cementing materials, analogous to lignin in timber, were however rather slow in hardening and developed little strength, so the resulting conglomerates tended to be weak. They did however possess one tremendous facility — they could be easily moulded at normal temperatures into any shape while still in the unset condition.

The amalgamation of this far-reaching property with the additional ones of high strength and stiffness must have exercised the minds of men for centuries. It was not however until the latter half of the eighteenth century that substantial improvements in the strength of the cementing medium began to be made. The advantages of using hydraulic limes which contained, in addition to calcium oxide, aluminates and silicates of calcium were noted in 1756 by John Smeaton in connection with the construction of the Eddystone Lighthouse and a little later, in 1796, James Parker patented what was known as 'Roman cement' produced by burning certain argillaceous limestones; the use of this material grew rapidly during the early nineteenth century.

It was a Leeds bricklayer, Joseph Aspdin, who in 1824 made the first step towards the production of the higher strength cements which are used today.[1-3] Aspdin's cement was composed of a mixture of clay and chalk which was burnt in a kiln and then ground to a powder, and, although not of exceptional strength, was used in some major works including the Thames Tunnel of Marc Isambard Brunel, the father of the perhaps better known Isambard Kingdom Brunel. Using Aspdin's basic ingredients, but by burning the mixture until the constituents began to fuse together, Isaac Johnson in 1845 produced a cement with strength characteristics vastly superior to any previously developed. Because upon hardening the new material resembled stone from the quarries near Portland, it was given the name 'Portland cement'.

The availability of Portland cement led to the production of one of today's most universal structural materials, concrete. Concrete is a mixture of various types and sizes of stone aggregate, sand and Portland cement, which upon addition of water hardens into an exceptionally strong and rigid material by the bonding action of the cement, and the matrix so formed exhibits a marked similarity to the structure of steel. The basic difference between the two matrices is one of dimension, the sizes of the individual component parts of the concrete mass being much larger than those of the steel, and in this sense, ordinary 'plain' concrete may perhaps be considered to have the attributes of a homogeneous material. The exceptional ease with which concrete can be moulded to any shape, or poured to form massive structures such as dams, has meant that it has become perhaps the predominant structural material of the twentieth century. The constraints imposed on architecture by the necessity of having to conform to a predetermined geometry, for instance that of a steel section, have been eliminated by the flexibility of concrete which allows for greater freedom and expression in design as can be noted perhaps in some of the beautiful plate and shell structures produced during recent decades.

There is however one major property that steel possesses that is absent from concrete. Like cast iron before it, it is a material that, although strong in compression, has a very low tensile strength. If, however, steel bars of high tensile strength are bonded into the concrete in zones that are subjected to tension, then these tensile stresses are resisted by the steel. The material thus formed is known as 'reinforced concrete', a true composite where distinct materials are combined in a non-random manner to produce overall structural characteristics superior to those of the individual components. Reinforced concrete, together with its later variant 'prestressed concrete', has been used extensively this century and it would be difficult now to find a single modern construction where this composite was not being used, such has been the impact of this versatile material.

The widespread use of concrete is living proof of the success that can be obtained from composites, and thus it is not surprising that investigations into materials of a similar mouldable form should have been considered.

Until the beginning of the present century, the materials developed, manufactured and used, whether homogeneous or composite, were basically inorganic in nature. Complex organic substances such as coal and oil were subjected to destructive processes to produce simpler chemicals such as coal gas and gasoline. However, during this century, organic chemists have developed the means of reversing this destructive process and of creating from the by-products materials that do not occur naturally. Most important among these new substances are the 'super-polymers', commonly called 'plastics', a term which in many cases is misleading, and the production of these materials has increased dramatically since the Second World War.

One class of super-polymers is produced by the addition to a basic chemical unit, known as a 'monomer', of a succession of similar units. For instance by repeating the monomer ethylene, C_2H_4, the substance polyethylene, or polythene, $(C_2H_4)_n$, is produced (Fig. 1.1a), this being known as a 'polymer'. It is seen that in this process the double valency bonds of the monomer are broken to form single bonds, the new form of the monomer being then referred to as a 'mer'.[4,5]

The replacement of one or more of the hydrogen atoms in ethylene by another atom or group of atoms produces a new group of compounds

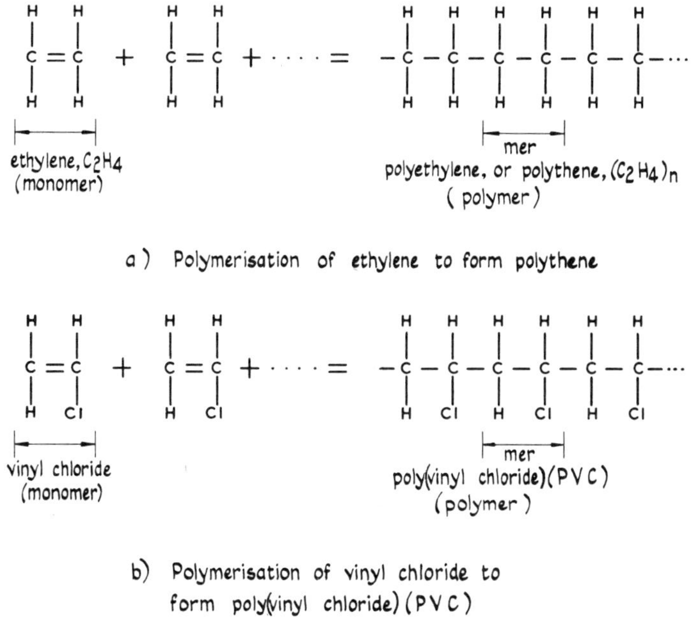

a) Polymerisation of ethylene to form polythene

b) Polymerisation of vinyl chloride to form poly(vinyl chloride) (PVC)

FIG. 1.1. Examples of addition polymerisation.

known as 'vinyls' which may be similarly polymerised. Thus the polymerisation of, for instance, the monomer vinyl chloride produces the polymer poly(vinyl chloride), commonly abbreviated to PVC (Fig. 1.1b), while the more complex vinyl monomer styrene when polymerised produces polystyrene.

This process of continually adding a repeating unit is known as 'addition polymerisation', and the resulting polymers because of their chain-like nature (Fig. 1.2a) become soft when heated and harden when cooled and are therefore termed 'thermoplastics'.

a) Chain-like molecules of a thermoplastic

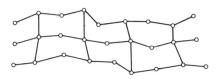

b) Cross-linked structure of a thermoset

FIG. 1.2. Basic structural forms of thermoplastic and thermosetting polymers.

There can be initiated, however, a further process known as 'condensation polymerisation' which may allow not only chains to be formed but also these chains to be linked together (Fig. 1.2b), thus producing a more rigid structure that has a greater resistance to temperature effects. The super-polymers in which this additional type of bonding occurs are known as 'thermosets' as they set hard and cannot be reformed by heating.

One of the earliest thermosetting plastics was produced by the Belgian chemist, Leo Baekeland, in 1907 from phenol and formaldehyde, and was known as 'bakelite'. Today this group comprises such materials as the epoxies, silicones and polyesters.

It is this second group of plastics, especially the polyester resins, that has been most frequently used in structural engineering applications. As materials in themselves plastics are unsuitable for major structural applications in which high loads or long spans are parameters since their stiffness and ultimate strength values are relatively low and their ability to creep is not negligible. These adverse properties are, of course, analogous to the adverse effect of the low tensile strength of plain concrete which makes it unsuitable for applications in which tensile stress is developed. Thus for the same reason that plain concrete must be 'reinforced' to produce a material with sufficient tensile strength, so plastics must be similarly reinforced if they are to provide materials with adequate structural properties. The reinforcement most commonly used in plastics, because of its good bonding characteristics with the resin and its high strength and stiffness properties, is glass fibre. It is very probable that carbon or boron fibres produce superior structural materials, but the cost of producing these restricts their use to more specialist applications. Plastics reinforced with glass fibre are however being used successfully in an increasing number of structural projects ranging from simple cladding panels to large sophisticated roof structures such as those at Dubai Airport and at Morpeth School in Stepney, London.[6]

Thus these synthetic composites have taken their rightful place in the family of structural materials. To steel, aluminium, and reinforced and prestressed concrete are now added reinforced plastics, the most common of which are the glass-reinforced polyesters known commonly as GRP, lightweight, non-corrosive, mouldable materials, the applications of which include large lightly loaded spans, pipes and containers.

It is with the structural properties of these materials that most of this book is concerned.

REFERENCES

1. MURDOCK, L. J. and BROOK, K. M., *Concrete Materials and Practice*, 5th edn., 1979, Edward Arnold, London.
2. PANNELL, J. P. M., *Materials of Civil Engineering*, 1957, Hutchinson, London.
3. SMITH, R. C., *Materials of Construction*, 2nd edn., 1973, McGraw-Hill, New York.
4. HIGGINS, R. A., *Properties of Engineering Materials*, 1977, Hodder and Stoughton, London.
5. VAN VLACK, L. H., *Elements of Materials Science*, 2nd edn., 1964, Addison-Wesley, Reading, Mass.
6. HOLLAWAY, L., *Glass Reinforced Plastics in Construction*, 1978, Surrey University Press, Glasgow.

CHAPTER 2

The Manufacture and Properties of the Components of Glass-reinforced Polyesters

2.1 THE POLYESTER MATRIX

Polyester resins, being thermosets, have the advantage over the thermoplastics in that they are more resistant to temperature changes and creep behaviour, and are preferred to their fellow thermosets, such as the epoxy resins, in that they are less expensive to produce, although the epoxies and other resins may be preferred purely from a materials viewpoint. To understand these materials and their relationship to other plastics, it is necessary to examine their composition, structure and properties.

2.1.1 The Basic Chemistry of Polyester Resins
The process of *addition* polymerisation, of which the fundamental example is the synthesis of polythene from the basic substance ethylene, produces long chain molecules, the presence of which is the basic characteristic of all polymers.

A molecule of ethylene, C_2H_4, may be represented structurally as shown in Fig. 2.1. In this molecule it is seen that to produce the stable electronic configurations of two and eight in the outer orbits of the hydrogen and carbon respectively, the hydrogen atoms must be connected to the atoms of carbon by 'single' valency bonds, whereas it is necessary for the carbon atoms themselves to be linked by 'double' bonds. Molecules possessing this multiple carbon to carbon bonding are said to be 'unsaturated', and such a bond is relatively simple to break chemically.

If the double bond in ethylene is broken, then large numbers of the resulting units, or 'mers', can form and link together with single bonds to

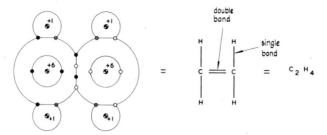

FIG. 2.1. Structural representation of a molecule of ethylene.

form the long chain molecule polyethylene, or polythene (Fig. 2.2). It can thus be seen that this breaking of the double bond produces a single bonded or 'saturated' chain molecule, and this process is characteristic of addition polymerisation. The length of the chains determines the viscosity of the resulting polymer. The longer the chains the greater are the intermolecular, or van der Waals, forces between the molecules and hence the more viscous the resulting material. Very long-chained molecules thus produce materials that can be regarded as solid. The application of heat increases the energy in the molecules so that the van der

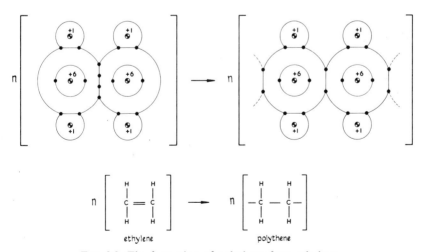

FIG. 2.2. The formation of polythene from ethylene.

Waals forces are overcome, thus reducing the viscosity of the material and causing it to soften. Hence this form of polymer is termed *thermoplastic*.

In this addition polymerisation process, there is one chemically exceptional characteristic — no material by-product is produced. In a vast number of chemical reactions, however, two or more products are created. In inorganic chemistry, one of the most fundamental of these is illustrated in the general equation

$$\text{base} + \text{acid} \rightarrow \text{salt} + \text{water}$$

For instance the addition of hydrochloric acid to the base sodium hydroxide produces sodium chloride (common salt) together with water. This may be expressed by the chemical equation

$$NaOH + HCl \rightarrow NaCl + H_2O$$

An analogous reaction takes place with the corresponding organic (or carbon based) molecules — that is, an organic acid when added to an organic base (an alcohol) produces an organic salt (an ester) together with water as a by-product. For instance, the reaction of acetic acid, $CH_3 \cdot COOH$, with ethyl alcohol, $C_2H_5 \cdot OH$, produces the ester ethyl acetate, $C_2H_5 \cdot COO \cdot CH_3$, and water:

$$C_2H_5 \cdot OH + CH_3 \cdot COOH \rightarrow C_2H_5 \cdot COO \cdot CH_3 + H_2O$$

Such a process, which is described in section 2.1.2, is known as a *condensation* reaction. In this example, it is noted that the reaction is between a *monobasic* acid (one COOH group) and a *monohydric* alcohol (one OH group), and that a simple ester results (Fig. 2.3a).

If now a polybasic acid, for example the dibasic acid adipic acid, $HOOC \cdot (CH_2)_4 \cdot COOH$, is reacted with a polyhydric alcohol, such as the dihydric alcohol ethylene glycol, $HO \cdot CH_2CH_2 \cdot OH$, then a multiple condensation process occurs and a long chain molecule is formed (Fig. 2.3b). This resulting molecule is known as a *polyester*.[1]

There are two forms of polyester, namely saturated and unsaturated. Saturated forms are produced from acids and alcohols (usually dibasic acids and dihydric alcohols, the latter often known as glycols) which are themselves saturated, that is they exhibit only singly bonded carbon atoms in their structure. The example above is of the formation of such a saturated polyester.

On the other hand, however, if either the acid or the alcohol is unsaturated, i.e. doubly bonded carbon atoms are present, then an

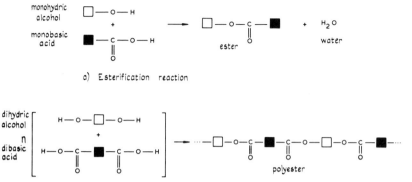

a) Esterification reaction

b) Polyesterification reaction

FIG. 2.3. Ester and polyester formation.

unsaturated polyester is obtained. In practice the acids are usually the unsaturated components, these being easier and cheaper to produce than unsaturated alcohols.

Such unsaturated acids that are predominantly used are fumaric acid and the anhydride form of its isomer, maleic acid (Fig. 2.4).

The difference between the saturated and unsaturated polyesters lies in the double bonds between carbon atoms, which as previously stated are relatively simple to break. In saturated polyesters, the carbon atoms are all singly bonded and hence the chain molecules are not amenable to change. Hence these polyester forms tend to be flexible and to show similar properties to those polymers produced by addition polymerisation. In the unsaturated polyesters, however, by the breaking of the

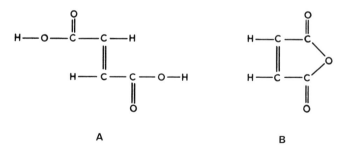

FIG. 2.4. Fumaric acid (A) and maleic anhydride (B).

double bonds, cross-linking of adjacent polyester chains through an unsaturated bond in an added monomer can be obtained, thus producing a material that is highly resistant to temperature change and therefore termed a *thermoset*.

The monomer most widely used to develop the cross-linking is styrene,[2] and Fig. 2.5 shows the cross-linking of the polyester chains produced from ethylene glycol and fumaric acid with this monomer. Other glycols may be used, the commonest perhaps being propylene glycol, $HO \cdot C_3H_6 \cdot OH$, which produces a polyester with less tendency to crystallise than that produced with ethylene glycol.[1]

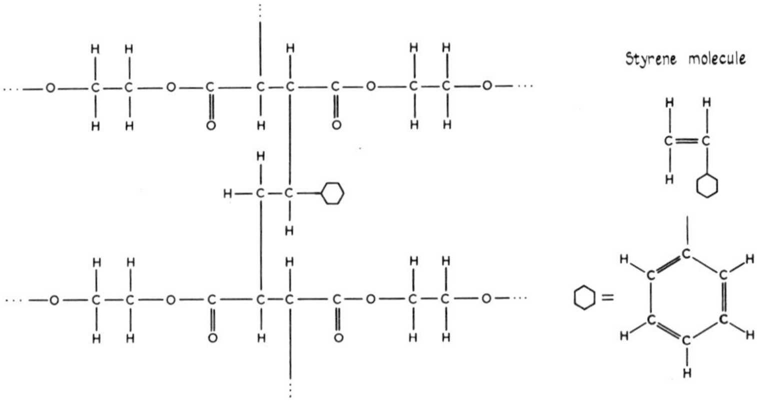

FIG. 2.5. Cross-linking of polyester chains with styrene.

The use of only an unsaturated acid produces such a high degree of cross-linking that the resulting resin tends to be rather brittle, so it is usual for polyester resins to contain at least two dibasic acid components, one being saturated, such as phthalic anhydride or adipic acid, and the other unsaturated, so that by suitably proportioning the quantities of each a degree of ductility can be introduced into the product.[2]

2.1.2 The Production of Unsaturated Polyester Resins

The manufacture of unsaturated polyester resins is carried out in three basic steps:

(a) the condensation reaction,
(b) the blending with the monomer, usually styrene,
(c) the curing process.

Because of the sensitivity of polyester resins to contamination in the manufacturing process special consideration must be given to the materials used for the processing equipment.

2.1.2.1 THE CONDENSATION REACTION

This reaction generally takes place in a stainless steel vessel which is fitted with an agitator to ensure a uniform product, and a condenser/receiver to collect the water produced by the reaction.

The liquid constituents are introduced first and the temperature raised to between 80 and 90°C, before any solid acid anhydride components are added, stirring taking place continuously.

In order to prevent cross-linking through the unsaturated acid component, which could occur at temperatures of the order of 200°C, an inert gas is introduced into the reaction vessel before the temperature is raised over several hours to about 200°C. This temperature is then maintained until the required degree of polyesterification (measured by the amount of unreacted acid remaining) is reached, after which the temperature is reduced.

2.1.2.2 THE BLENDING WITH THE MONOMER

The addition of the monomer, usually styrene, to the product of the condensation reaction would, in the presence of heat or light, initiate the cross-linking process and produce the solid polyester resin. To prevent this occurring, an 'inhibitor' must be added to the base polyester prior to addition of the monomer.

These inhibitors react with any activated double bonds so that further polymerisation is prevented. Thus they maintain the resin in a liquid form, and allow for long periods of storage, if this is at temperatures low enough to minimise double bond activation.

After the addition of the inhibitor the base polyester is blended with the monomer, this process being effected at about 50°C to 65°C. Any special additives required, such as room temperature accelerators, are also included at this stage.

2.1.2.3 THE CURING PROCESS

The processes described previously are the manufacturing procedures required to produce the polyester resin in a bulk liquid form. The final process of converting the liquid form into a solid state involves the addition polymerisation reaction whereby the blended monomer reacts with the unsaturated groups in the polyester chains to produce the cross-linked thermoset. This process is known as 'curing'.

It is customary to initiate curing by the addition of an 'initiator' to the liquid form, the most common form of initiator being the organic peroxides.[1]

The room temperature accelerators introduced into the resin during the blending stage either serve to increase the rate of performance of the initiator or enable the initiator to become effective at temperatures lower than those necessary if the initiator alone were used.

2.1.3 The Properties of Cured Unsaturated Polyester Resins

The properties of any particular cured polyester resin will obviously be dependent upon the nature, quality and quantity of its constituents, and on the control exercised in the manufacturing process. In this sense the production of resin is analogous to the manufacture of concrete, a material whose properties vary with the same parameters. Hence the properties of polyester resin, although they may be described extensively in a qualitative manner, can be quantified only over a range of values, and any particular resin may have a number of desirable properties while at the same time exhibiting certain unwelcome features.

From a structural viewpoint, the most important properties of the material are its basic mechanical characteristics of specific gravity and shrinkage, strength and stiffness, and its creep behaviour, but its resistance to environmental effects and to fire are also important parameters for consideration.

These basic properties are now considered in turn.

2.1.3.1 SPECIFIC GRAVITY AND SHRINKAGE

In comparison with most other structural materials, cured polyesters have very low specific gravities, having values of the order of 1·1 to 1·4,[3] i.e. densities between 1100 and 1400 kg/m^3. Although most timbers have specific gravities less than unity, being less dense than water, the other common structural materials have far higher values. For instance, structural steels have a specific gravity of about 7·8, aluminium alloys 2·7, and concrete of the order of 2·4.

A shrinkage of the order of 5–10% by volume can occur during the curing of polyester resins, a feature that can be rather undesirable. The shrinkage of their relatives, the epoxy resins, on the other hand is much smaller, being of the order of 2% by volume.

2.1.3.2 STRENGTH AND STIFFNESS

Unlike structural materials such as steel and other metals, and even concrete, the strength and stiffness properties of plastics are in general more sensitive to changes that can occur during normal serivce, such as fluctuations in temperature and variations in the rate of loading. Hence data obtained using the same testing procedures as for other materials could be extremely misleading, so to construct a realistic picture of these properties for plastics, in general a series of tests are necessary over a range of temperatures and also at various rates of loading.

In polyester resins, however, and in the other highly cross-linked thermosetting plastics, although increase of temperature does cause a deterioration in ultimate strength and stiffness, the resistance to temperature effects is fairly substantial over normal service ranges; also it is found that except for very high straining rates, such as impact, any effect of the rate of loading may be neglected.[4] This latter conclusion is most useful since in structural practice the rate at which load application takes place is almost impossible to quantify.

Although the properties of cured polyester resin are dependent on the ingredients selected, the stress–strain ratio in both tension and compression may generally be assumed constant over the working range of stress. This ratio is a measure of the material stiffness and is known as Young's modulus or the modulus of elasticity; comparison of this property for cured polyester resin shows that it is far lower than that for other structural materials. For instance, whereas steel has a modulus of elasticity of about $210 \, kN/mm^2$, aluminium alloy $70 \, kN/mm^2$, and concrete a value within the range 15–$35 \, kN/mm^2$, the short term value for cured polyester resin at ambient temperatures may only be in the region 2.5–$4.0 \, kN/mm^2$, this being less than the value for structural timbers.[5] The further material property of Poisson's ratio (the significance of which will be discussed in Chapter 4) for cured polyester resin is of the same order as for metals, lying in the range between 0.3 and 0.4. In comparing the short term ultimate tensile and compressive strengths of various materials, however, it is observed that the cured polyester resins fare much better. Ultimate tensile strengths of the order of 45–$65 \, N/mm^2$ and ultimate compressive strengths of about twice these values can be obtained, which, while lower than the tensile strengths of most materials, compare quite well on a compressive strength basis. It should be noted, however, that together with all other plastics, polyester resins exhibit a reduction in ultimate strength with time, although the rate of reduction is not as great as in the thermoplastics.

Typical properties of the various structural materials are compared in Table 2.1.

TABLE 2.1
COMPARISON OF THE APPROXIMATE STRUCTURAL PROPERTIES OF VARIOUS MATERIALS

Material	Specific gravity	Ultimate tensile strength, N/mm^2	Ultimate compressive strength, N/mm^2	Modulus of elasticity in tension or compression, kN/mm^2	Poisson's ratio
Polyester resin	1·1–1·4	45–65	90–130	2·5–4·0	0·3–0·4
Structural steels	7·8	400–550	—	210	0·3
Structural aluminium alloys	2·8	200–450	—	70	0·3
Concrete	2·4	3	40	15–35	0·1–0·3
Douglas fir (parallel to grain)	0·5	140	50	10	—

Values quoted are those obtained from short term tests at normal atmospheric temperatures.

2.1.3.3 Creep Behaviour

Of the traditional structural materials, concrete is the only medium that exhibits any appreciable creep behaviour, under common environmental conditions, and this has been the subject of extensive investigations. The creep in plastics, especially in the thermoplastics, is far greater and is dependent both on the temperature and on the magnitude of the applied stress.

The effect of continued application of stress on polymers, such as polyester resin, is to cause a straining of the molecular bonds. This molecular response is slow to reach an equilibrium state and so the material continues to deform for long periods after application of the load. Upon removal of load, some non-elastic behaviour of the material may be noticed. Likewise, as the temperature is increased, so the molecular activity is magnified, and additional creep behaviour ensues.

The creep rate can be reduced by a decrease of molecular movement and this can be partially achieved in the thermosetting plastics by ensuring full curing of the resin.

2.1.3.4 Environmental Effects

In addition to a determination of the basic mechanical properties of the polyester resin, a knowledge of the long term effect of weathering on the material is also of value. The environmental conditions that cause weathering can be largely overcome by adequate protection of the resin with additives and coatings. In particular the deleterious effect of water is reduced by ensuring complete curing of the resin, and the effect of sunlight, in particular its ultraviolet component which causes untreated resins to yellow appreciably, can be almost eliminated by the addition of a small quantity of strong organic ultraviolet absorbers or light stabilisers such as the hydroxybenzophenone derivatives.[6] The application of a thin film of poly(vinyl fluoride) to the polyester surface has been shown to provide substantial protection against surface erosion and loss of colour,[7] which can also be caused by heat, and if required fire retardant additives, such as antimony oxide, can be dispersed in the resin.[8]

2.2 THE GLASS FIBRE REINFORCEMENT

A large number of materials, e.g. jute, asbestos, carbon and boron, has been used for the fibre reinforcement of the plastic matrix, the main function of the fibres being to carry the majority of the load applied to the composite and to improve the stiffness characteristics of the resin. However, the most widely used material for the reinforcement of polyester resin is glass fibre in all its various forms, partly because of its high strength and its low specific gravity, partly because of its chemical inertness, and partly because of its being relatively inexpensive to produce. Like the polyester resin, the glass fibre reinforcement can be composed of various materials in various quantities, so it is first necessary to consider of these various forms which are the most suitable, and also over what range of diameters and lengths the fibres should be formed.

2.2.1 The Structure and Types of Glass

Although glasses are commonly conceptualised as hard, brittle solids, they are in fact produced from a fusion of inorganic oxides which have been rapidly cooled so that the process of crystallisation is reduced in the formation of the hard substance, and may be considered to be supercooled liquids.

A solid that possesses such a non-crystalline structure is said to be *amorphous*, i.e. lacking a definite structure. How this property, which is common to all glasses, is produced may be simply described as follows.

The major component of glasses is quartz, a crystalline form of silicon dioxide or silica, SiO_2. In this crystalline state the basic unit may be considered to consist of the silicon atom singly bonded to four oxygen atoms at the corners of a tetrahedron as shown in Fig. 2.6. The apparent

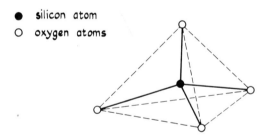

FIG. 2.6. Basic tetrahedral unit of quartz. ● = silicon atom, ○ = oxygen atoms.

lack of stable electronic outer orbit configurations in the oxygen atoms in this diagram is of course accounted for by noting that each of these atoms combines with the silicon atom of an adjacent tetrahedron, each oxygen atom being thus common to two tetrahedra. Combination of the tetrahedral units hence produces the structure shown, for simplicity in a two-dimensional form, in Fig. 2.7a.

In changing from the solid to the liquid state the orientation of the individual tetrahedra becomes more random as the molecular bonding becomes weaker and the structure more flexible. The quartz now becomes fused silica. Rapid cooling of the liquid makes the reversal to the regular lattice structure impossible and the new form inherits the amorphous network structure shown in Fig. 2.7b, when it is known as silica glass. The high strength of glass in a function of this network structure.

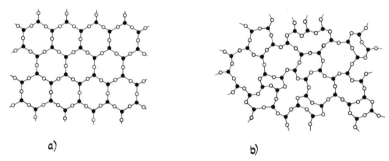

FIG. 2.7. Two-dimensional structural representations of (a) quartz and (b) silica glass.

Variation of glass properties can be attained by the addition to the silica of metallic oxides, in particular those of aluminium, calcium, magnesium, sodium and potassium. The effect of the extra oxygen supplied by these additives is to cause some loss of continuity in the network structure,[9] thus enabling a more workable glass to be produced, while the metallic ions produce other property variations, such as differences in the melting point and in resistance to deterioration.

For the purpose of fibre reinforcement, the most commonly used glass is a lime–alumina–borosilicate complex known as 'E' glass. This glass exhibits high strength and possesses good characteristics for fibre formation. A glass with a higher silica content and an absence of calcium, designated 'S' glass, has superior strength and stiffness characteristics to 'E' glass but its production costs are unfavourable. Other forms of glass are manufactured for specialist purposes. For example a relatively low strength, low stiffness glass known as 'D' glass has a low dielectric constant and is used in electronic applications.[10]

2.2.2 The Production of Glass Fibres

The ease with which production of glass fibres can be attained is dependent on the quality of the glass, in particular the absence of small solid particles and the exhibiting of adequate viscous properties being vital.

The basic process of fibre production is shown diagrammatically in Fig. 2.8. Refined glass, either molten or in the form of solid marbles of

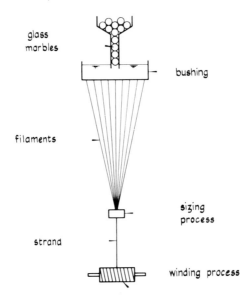

FIG. 2.8. Glass fibre production process.

about 20 mm diameter, is fed into a heated platinum or platinum-alloy 'bushing' containing usually a multiple of 204 orifices, each having a diameter in the range 0·75–3 mm. By means of a mechanical winder, the molten glass is pulled through the orifices in the 'bushing' at a speed of the order of 3000 m/min, thus producing a continuous filament from each orifice.[3] The diameter of these filaments may range from about $2·5 \times 10^{-3}$ to 20×10^{-3} mm depending on the size of orifice and the drawing speed and temperature.[10]

The filaments are then drawn together to form a strand, for in this form brittle fracture due to crack propagation emanating from surface flaws can be limited to single filaments and so failure of the complete strand is prevented. Thus a strand will have a greater fracture strength than a single fibre of the same total diameter, for in the latter case the development of a surface crack would cause brittle fracture of the complete specimen.

Of course the individual filaments would be mutually extremely abrasive if allowed to come into contact with each other without adequate protection, and their surfaces would be markedly flawed, thus leading to a substantial degradation of strength and possible fracture. Therefore before the strand is formed the filaments are treated with a size, usually a starch–oil emulsion, to protect their surfaces.

Finally the strand is wound on to a tube or drum.

2.2.3 The Properties of Glass Fibres

Many of the characteristics of glass in bulk or fibre form are similar, but its strength in the fibre state, partly due to the relative absence of surface flaws, is far higher than in the bulk form, although its mechanical resistance is lower. The chemical composition of the glass also affects the fibre properties, particularly the strength and stiffness values.

Unlike the polyester resin, glass fibres are extremely creep resistant in dry conditions and generally exhibit chemical inertness. Any effect of corrosion by water can be reduced by maintaining a low sodium and potassium content, and under the normal ambient temperatures experienced little effect on the glass is then observed.

Therefore the predominant properties of glass fibre that require quantification are its specific gravity, strength and stiffness.

2.2.3.1 Specific Gravity

The specific gravity of typical bulk 'E' glass is about 2·58 whilst that of the fibre form is a little less, being approximately 2·54. 'S' glass is slightly less dense, its average fibre form exhibiting a specific gravity of about 2·49.

The glass fibres are thus on average approximately twice as dense as the polyester resin.

2.2.3.2 Strength and Stiffness

Unlike most metals, glass is characterised by a linear elastic stress–strain relationship to failure, and in fibre form exhibits exceptionally high strength, far higher than any of the traditional constructional materials. 'E' glass, for instance, in fibre form exhibits an average tensile strength of about $3·45 \, kN/mm^2$ in the newly drawn and untouched condition, which is several times the strengths of the metals that are used structurally. In combination with a moderately low density, this strength results in an extremely high strength/weight ratio.

The modulus of elasticity, on the other hand, is not as exceptional, this property for 'E' glass being about 72.4 kN/mm^2, a value similar to that for aluminium.

The measured values of Poisson's ratio for glass fibres are variable but usually lie in the region of 0·2.

The proportions of the major components by weight, together with strength and stiffness values, for both 'E' and 'S' glass are shown in Table 2.2.

2.2.4 The Forms of Glass Fibre Reinforcement

Glass fibres are used in a variety of forms as reinforcement in plastics; the form of reinforcement will depend on the method of production of the final product, the properties required and the cost of the various forms of fibre reinforcement.

Basically there are three forms which the reinforcement may take, namely

(a) chopped strand mat,
(b) rovings and woven rovings, and
(c) woven fabrics

2.2.4.1 Chopped Strand Mat

This form of glass fibre reinforcement is manufactured from glass strands cut into approximately 6–50 mm lengths. The process results in a random mixture of portions of strands and individual filaments, the proportion of each being determined by the amount and nature of the size that has been used in the initial strand forming process. The more effective the binding property of the size the greater the proportion of strand portions to filaments.

The strand–filament mixture is then bonded together with a suitable resin binder, thus forming a random two-dimensional fibre mat, the strength of which will be determined by the quantity of strand and filament portions present.

This form of reinforcement is generally used, except for applications where high strength is required. In these cases the reinforcement would probably take the form of rovings or woven fabrics.

TABLE 2.2
COMPOSITION AND PROPERTIES OF 'E' AND 'S' GLASS

Glass	Composition by wt of major components					Ultimate tensile strength, kN/mm^2	Modulus of elasticity in tension or compression, kN/mm^2
	SiO_2	Al_2O_3	CaO	MgO	B_2O_3		
'E' glass	54	14	17·5	4·5	10	3·45	72·4
'S' glass	65	25	0	10	0	4·60	85·5

Small quantities of sodium, potassium and iron may be present in these glasses.

2.2.4.2 Rovings and Woven Rovings

Rovings consist of a group of continuous, parallel, untwisted strands, bundled together, the number of strands varying from 6 to 150, although the most common number is 60.

This form of reinforcement exhibits very high strength characteristics in the direction of the strands and is ideally used in the reinforcement of high-performance closed structures such as pipes, tanks and rocket casings.

In addition to supplying high unidirectional strength, rovings may be woven to produce high strengths in two directions at right angles to each other, the relative strengths in the two directions being functions of the volume of the rovings in each direction.

2.2.4.3 Woven Fabrics

This form of reinforcement is expensive and used only where very high strength composites are required. In these fabrics the strands are twisted to form *yarns* which are then formed into uniform close-woven cloths with various forms of weave. Because of the fine weaving processes employed, a denser reinforcement can be obtained with these fabrics than with chopped strand mat or rovings.

In addition to these three main forms of reinforcement there are also other minor forms such as surface tissue and overlay mat. These are very thin tissues of fibre able to absorb a large proportion of resin and are used when surfaces of high resin content are required.[4]

To see how these various forms of reinforcement are satisfactorily combined with the polyester resin to form glass fibre composites suitable for structural use is the purpose of the next chapter.

REFERENCES

1. LAWRENCE, J. R., *Polyester Resins*, 1960, Van Nostrand Reinhold, New York.
2. DIAMANT, R. M. E., *The Chemistry of Building Materials*, 1970, Business Books, London.
3. BROUTMAN, L. J. and KROCK, R. H., *Modern Composite Materials*, 1967, Addison-Wesley, Reading, Mass.
4. BENJAMIN, B. S., *Structural Design with Plastics*, 2nd edn., 1982, Van Nostrand Reinhold, New York.
5. HOLLAWAY, L., *Glass Reinforced Plastics in Construction*, 1978, Surrey University Press, Glasgow.

6. PARKYN, B., LAMB, F. and CLIFTON, B. V., *Polyesters, Vol. 2, Unsaturated Polyesters and Polyester Plasticisers*, 1967, Iliffe Books, London.
7. GUMERMAN, C. and MACKAY, G. R., Evaluation of polyvinylfluoride film as a weatherable finish on reinforced polyesters, *Proc. 18th Technical and Management Conference, Reinforced Plastics Division, SPI*, Feb. 1963, New York.
8. BOENIG, H. V., *Unsaturated Polyesters: Structure and Properties*, 1964, Elsevier, Amsterdam.
9. HIGGINS, R. A., *Properties of Engineering Materials*, 1977, Hodder and Stoughton, London.
10. VINSON, J. R. and CHOU, T. W., *Composite Materials and their Use in Structures*, 1975, Applied Science Publishers, London.

CHAPTER 3

The Manufacture and Characteristics of Glass-reinforced Polyesters

3.1 THE METHODS OF MANUFACTURE

The application of fibre-reinforced materials now extends over a broad spectrum of technology, the components manufactured in the new composite materials embracing a wide range of size, shape and property requirements. Because of this diversity in application, several component manufacturing processes have been developed.

The particular process used for manufacture will depend primarily on the geometry and size of the product and in what situations the product is to be used. It will also depend on the nature of the reinforcement and form of resin used, on the number of similar articles that are to be produced, and on the cost, either in labour or in machinery.

Basically there are two groups of systems used for the fabrication of glass fibre-reinforced polyester products, namely

(1) open mould processes,
(2) closed mould processes.

The open, or single mould, processes are those that are generally used for the production of the large components for structural engineering applications, while the closed mould systems are predominantly used for smaller components, usually with more precise geometry, for applications in other branches of technology. The various processes comprising both these groups are depicted in Fig. 3.1, and each will now be considered.

3.1.1 Open Mould Processes
As the name implies, these processes use only one mould, and hence only the surface of the product in contact with the mould is able to attain a

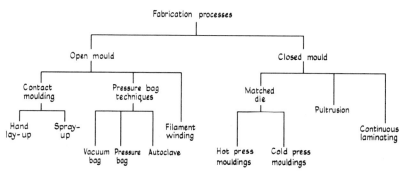

FIG. 3.1. Processes used for the manufacture of GRP products.

smooth finish. The methods also utilise two advantageous characteristics of polyester resin, namely that in standard environments neither heat nor pressure is essential to attain complete curing, although at times their application may supplement the natural cure.

This group of processes can be further subdivided into

(1) contact moulding,
(2) pressure bag techniques,
(3) filament winding,

the second group lying somewhere between the open and closed mould processes, and the third technique being mainly applicable to components exhibiting a surface of revolution.

3.1.1.1 CONTACT MOULDING

(a) Hand Lay-up
This method, the oldest, simplest, and most common technique for the production of components in fibre-reinforced plastics, is ideally suited to the production of a small number of similar components such as commonly occur in structural engineering practice.

Using low-cost equipment and having little restriction on size, the method exhibits a direct parallel to the techniques for producing precast reinforced concrete components.

The mould, which may be constructed of a number of materials although perhaps the most common is GRP itself, is first treated with a suitable release agent, such as a mineral oil, to ensure that the final product can be easily removed from the mould without damage. When this has been allowed to dry, an even layer of resin up to about 0·5 mm thick, often containing additives and pigments to give weathering resistance and colour to the surface, and sometimes reinforced with a surface tissue mat, is applied to the mould surface and allowed to cure partially.[1,2] This resin layer, known as the 'gel coat', serves two purposes. Firstly it protects the subsequent glass fibre layers from deleterious environmental effects, and secondly it ensures that the mould surface is exactly reproduced in the finished component.

When cured sufficiently, resin is brushed on to the 'gel coat' and the first layer of glass fibre reinforcement, which may be in the form of chopped strand mat, woven roving or woven fabric, thoroughly impregnated with the resin to ensure that any trapped air is removed, and laid in position with a brush or roller. Further layers of resin and reinforcement are then added until the required thickness of the laminate is obtained (Fig. 3.2a).

Being relatively labour intensive, the success of the technique and the quality of the finished product are highly dependent upon the skill of the

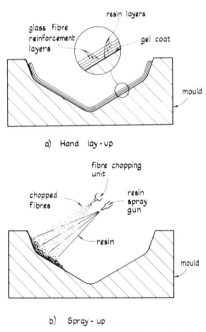

FIG. 3.2. Contact moulding methods.

fabricator. However, the process is extremely flexible, allowing design alterations to be easily made, and capable of producing composites with glass contents normally between 30 and 50% by weight.

(b) Spray-up
In essence the spray-up technique is similar to the hand lay-up method, except that it is a little less labour intensive and involves rather higher equipment costs.

The technique involves the simultaneous deposition of chopped glass fibre roving and resin on to the mould by means of a spray gun (Fig. 3.2b). The roving, which is the cheapest form of glass fibre reinforcement, is fed through a chopping unit and projected into the resin stream. The glass–resin mixture is then rolled by hand to remove any air that may be present in the composite.

The advantages of the method lie in the reduction of labour costs and in the ease with which complex shapes can be formed. However, considerable expertise is required by the operator in maintaining a uniform glass/resin ratio throughout and in controlling the thickness of the composite; because of these factors sprayed-up composites are likely to show more variation in quality than those produced by hand lay-up techniques. Composites containing about 30% by weight of glass reinforcement can be formed by the process.

3.1.1.2 Pressure Bag Techniques
These processes, as the name implies, are further developments of the contact moulding methods in which pressure is exerted on the open face of the moulding so that (i) there is a greater assurance of the product being free from voids, (ii) a higher proportion of glass reinforcement can be accommodated in the composite, and (iii) the open face is of a higher quality than can be attained by ordinary contact moulding.

Pressure bag methods are extremely practical techniques for fabricating closed vessels without incurring the cost of sophisticated mechanisation, and three common forms of such methods are now described.

(a) Vacuum Bag
In this technique a normal contact moulding is first carried out, and before curing has advanced a rubber membrane is placed over the component, all joints are sealed, and a vacuum is created under the bag (Fig. 3.3a). This results in a pressure of up to 1 atmosphere being applied to the surface of the moulding, thus forcing out air and excess resin and

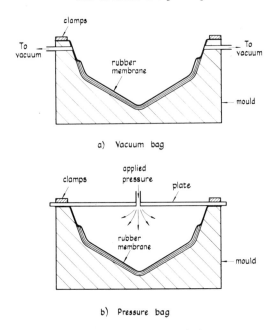

FIG. 3.3. Pressure bag techniques.

causing glass contents of up to approximately 55% by weight to ensue. Because the resin has a deleterious effect upon the rubber forming the bag, a protective sheet of cellophane is placed on the laid up composite prior to application of the bag.[3]

(b) Pressure Bag
This process is similar to the vacuum bag method except that pressure is applied directly to the open surface (Fig. 3.3b). After normal contact moulding and application of the cellophane and rubber bag, the mould is sealed with a plate and pressurised up to about $350 \, kN/m^2$, which is a little over 3 atmospheres. Because of the higher pressures utilised as compared with the vacuum bag process, the glass content of the composite can be increased to about 65% by weight, with a corresponding increase in confidence in the elimination of voids.

Although superior composites can result from the process, more equipment is required than with the more basic contact moulding methods, and hence it is a more costly technique. Furthermore, the nature of the process limits its application to those products having a convex surface in contact with the mould.

(c) *Autoclave*
A further modification of the pressure bag method is the application of pressure within an autoclave where pressures up to $700\,kN/m^2$ can be applied. Naturally, an increase in quality is obtained — up to about 70% by weight of glass can be introduced into the composite — but at the same time greater equipment costs are incurred, and also the size of the products is limited by the magnitude of the autoclave itself.

3.1.1.3 FILAMENT WINDING
Although the pressure bag processes can be used for the manufacture of closed vessels, when such structures are axisymmetrical and are required to possess exceptionally high strengths, for instance in rocket casings, pressure vessels or pipes,[4] the highly mechanised method of filament winding[5] becomes a viable technique (Fig. 3.4).

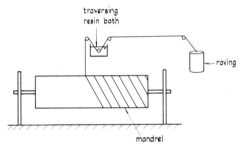

FIG. 3.4. Filament winding.

This process, necessarily more expensive than those so far considered, uses continuous reinforcement, usually roving, thus utilising the inherent strength of glass fibre most effectually. This continuous reinforcement is either fed through a resin bath, or if resin-preimpregnated reinforcement is used, passed over a hot roller until tacky, and then wound on to a rotating mandrel via a traversing mechanism.

The reinforcement may be laid at any helix angle by suitably adjusting the movement of the traversing mechanism relative to the mandrel. Indeed, the wall of a finished product may consist of several fibre layers in various orientations to satisfy the strength requirements of the particular design.

After the composite has cured on the mandrel, either at ambient temperature or in an oven, the mandrel is removed from the product. In the case of open-ended components such as pipes, the mandrel may be constructed from steel and simply forced out, but for the manufacture of closed vessels, the mandrel must be constructed from a low melting-point metal or water-soluble salt that can be removed from the finished component by melting or in solution.[6,7]

Although utilising expensive machinery, and being restricted to fairly easily wound shapes, the filament winding process produces extremely strong, high quality composites capable of containing up to 90% of glass by weight, a proportion higher than that produced by any other manufacturing technique.

3.1.2 Closed Mould Processes

These processes, in which the product is formed within a closed space by two moulds, are essential in the rapid continuous mass production of precision fibre-reinforced articles. The curing process is often accelerated by the application of heat to the composite, and unlike the open mould processes, these techniques can ensure that all surfaces of the component are manufactured to a predetermined form of high quality. In addition, the inherent application of pressure in the process contributes substantially to the elimination of voids in the composite.

Because of the high equipment costs necessary before production can commence, these methods would be uneconomical for the manufacture of only a small number of items, and are therefore inherently designed for large production runs where little skilled labour is required.

Depending upon the nature of the manufactured product, the processes may be considered in three forms, namely

(1) matched die processes,
(2) pultrusion process,
(3) continuous laminating.

The first of these forms finds application in the mass manufacture of relatively small articles of complex shape requiring all-round high quality surfaces, such as small boat hulls, car bodies and safety helmets, whilst the second and third forms are used for the continuous production of high quality composites in the form of rods or sheets respectively.

3.1.2.1 MATCHED DIE PROCESSES
Due to the geometrical complexities of components produced from

matched dies, the reinforcement is generally of chopped fibres in one of two forms. Either the chopped fibres are bonded together in the form of a mat conforming to the shape of the article, this mat being known as a 'preform', or they are simply mixed into the resin, thus forming a 'premix' or 'dough moulding compound'.

The latter form of composite exhibits greater discontinuities in the reinforcement than those involving a 'preform', and hence flows far more readily during the formation of the component. However, because of this discontinuity, the glass content tends to be low, perhaps only 25% by weight being attained compared with possibly 50% by weight in 'preform' composites, and also the strengths achieved with a 'premix' are lower than those obtained in 'preform' mouldings. Matched die processes may be executed with or without the application of heat, as is now briefly described.

(a) Hot Press Mouldings

The hot press moulding techniques consist in confining hot-curing resin and glass fibre reinforcement, either in 'preform' or 'premix' form, between polished matched metal dies which are brought together under pressure and heat. The dies may be manufactured from steel, cast iron or aluminium, the highest quality surfaces probably being obtained from dies of chrome-plated steel[1] (Fig. 3.5).

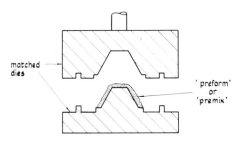

FIG. 3.5. Matched die process.

The pressures used in these processes may range from 400 to 4000 kN/m^2 and sometimes higher, with temperatures of the order of 120–150°C.

The application of heat enables rapid curing to occur, an essential ingredient in a system designed for mass production.

(b) *Cold Press Mouldings*

Essentially the same as the hot press technique, but because there is no application of heat the cold press moulding system can make use of plastics in the manufacture of the moulds, although this implies a restraint on the pressures that can be applied. While capital costs are therefore lower than those incurred in the hot press systems, curing times are higher and the geometrical precision obtained in the moulding may be slightly inferior.

3.1.2.2 PULTRUSION PROCESS

The production processes considered up till now have been those engineered for the manufacture of discontinuous components, although the matched die processes are designed for the mass production of such articles and in this respect may be considered to effect the continuous manufacture of a discontinuous product. The pultrusion process, on the other hand, is one of the techniques designed for the production of a continuous product.

Pultrusion is used for the production of continuous straight stock with a small, constant cross-section and hence has a parallel in the processes used in the production of steel sections. Indeed, the sections commonly used in steel construction, such as rod, angle, channel, and I-section, can all be produced in glass fibre composites by the pultrusion process, although the maximum cross-sectional dimension may be limited to about 150 mm.

Basically, the process consists of impregnating continuous glass strands, usually in the form of roving, in a resin bath before drawing them through a die to obtain the desired shape of the section (Fig. 3.6). Rapid curing is carried out in an oven or, alternatively, by externally heating the die.

A glass content of between 60 and 80% by weight can be achieved by

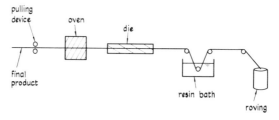

FIG. 3.6. Pultrusion process.

the process and because of the continuity of the reinforcement the end products, as in the filament winding process, possess exceptionally high strengths in the direction of the reinforcement.

3.1.2.3 CONTINUOUS LAMINATING

A process similar in kind to the pultrusion technique but designed to produce a composite in flat or corrugated sheet form is known as continuous laminating.

The reinforcement layers used for the process, usually fabric or chopped strand mat, are impregnated in resin and sandwiched between cellophane sheets before passing first between rollers to control the thickness of the laminate and then through a heating zone to effect the curing.

The role of the cellophane sheets is to assure that the surfaces of the laminate are perfectly smooth, this being attained by the shrinking and hence tensioning of the cellophane during the heating process.[2]

The width of the laminate is of course limited by the size of the rollers, although there is no such restraint on the length of the product. The laminate thickness in this technique is extremely uniform, but limited to about 3 mm, whilst the glass content may be of the order of 30% by weight.[8]

3.2 MATERIAL CHARACTERISTICS

The properties of materials composed of a resin matrix and glass fibre reinforcement are clearly dependent upon the nature, relative configurations and proportions of the individual constituents, and hence the range of variations and characteristics that such composites can possess is extensive.

Although numerous variations may occur in the properties of the various forms of composite, it is not surprising that due to the low specific gravity of the resin and the high strength of the glass fibres all forms of glass-reinforced composites possess an exceptionally high strength/weight ratio (measured in the direction of the reinforcement) compared with that for other structural materials. Indeed it is this characteristic that gives the materials their *raison d'être* and makes them so uniquely suited to applications where low weight is essential or desirable, such as in aerospace technology or in large span roof construction. The relatively high strengths that can be achieved in the medium are, however, significantly dependent on the quality of manufacture, especially in the elimination of air voids, the presence of which, even in small proportions, may cause a large reduction in strength.

While all forms of composites exhibit these superior strength characteristics when compared with the density of the material, the actual strength and deformation properties of a particular composite are highly dependent upon the nature and quantity of the reinforcement used, and thus these properties may be quite different from those of the traditional structural materials. The reason may be understood by considering the philosophy basic to fibre-reinforced composites. The underlying concept of the structural behaviour of such materials is that the great majority of the applied load is carried by the fibres, which are stronger and stiffer (although more brittle) than the resin matrix, while the resin itself provides the structural body and also stability to the glass reinforcement, especially in conditions where the material is subjected to compression. Traditional structural materials may usually be considered homogeneous and also 'isotropic' (i.e. they possess the same material properties in every direction at any point), a property which greatly simplifies the quantitative description of the material behaviour. Even in reinforced and prestressed concrete, which is composed of two materials, these properties may be assumed applicable owing to the fact that one material, namely concrete, is predominant. In contrast, in glass fibre-reinforced composites the proportions of the two constituent materials are of the same order and hence the assumption of homogeneity may not be valid, and the material properties will depend both on the form and on the number of layers of reinforcement used. Thus, for example, whilst because of the random nature of chopped strand mat, composites containing this form of reinforcement may be considered to be isotropic in the plane of the lamina, those reinforced with woven rovings or fabrics will exhibit different properties in various directions, and such materials are said to be 'anisotropic'. However, the nature of these latter forms does allow some simplification of a general anisotropic condition due to the fact that in the plane of a lamina reinforced with a single layer of woven roving or fabric there exist two mutually orthogonal directions of material symmetry, namely the two directions of the strands of the reinforcing medium. Such laminae are known as 'orthogonally anisotropic', a term usually abbreviated to 'orthotropic'. The mechanical behaviour of orthotropic, or more generally, anisotropic materials, as will be seen, is quite different from that of materials displaying isotropic properties.

Since the fibres are the main load-carrying component of glass fibre-reinforced composites the load–deformation characteristics of a composite are highly dependent on those of the constituent fibres, and thus linear stress–strain behaviour to failure is the usual assumption made in glass fibre material design. The loads applied to the composite are transmitted to the fibres both by shear through the resin matrix and by interfacial bonding between the resin and the glass. Hence because of the weaker nature of the resin, glass fibre-reinforced laminates may show inherent weaknesses in interlaminar shear characteristics. The necessity for an adequate resin-to-glass bond is thus obvious, and is obtained by adequate immersion of the fibres in the resin. Therefore whilst higher strengths can be obtained with higher proportions of glass, it must be ensured that there is enough resin content to procure adequate bond.

Besides showing quite different stiffness and strength properties from other materials, glass fibre-reinforced polyesters also display considerable creep behaviour, far greater than that which occurs in most other structural materials. This behaviour is due mainly to the creep properties of the resin, so creep resistance is generally increased by an increase in the fibre content. Even though the magnitude of creep behaviour may be reduced by the use of high fibre volumes, nevertheless it is almost always a factor which must be considered in any structural application.

Although environmental conditions such as fire, light and moisture[9-11] can have a detrimental effect upon the characteristics of glass-reinforced polyesters, such effects can be modified by the use of additives to the resin or by external coating.[2] The absorption of water can have particularly deleterious effects if not adequately inhibited as it can cause the resin to plasticise and possibly the resin–fibre interface to degrade. Therefore the need for a high degree of fibre–resin bonding is again seen to be essential. Such absorption can occur readily at temperatures in the region of 80°C, but less so at ambient conditions.

These then are some of the major characteristics of glass-reinforced polyesters which make them differ from other materials. It will be the purpose of the remainder of this book to examine the structural characteristics of such composites and to show how such materials may be used in specific design applications.

REFERENCES

1. HOLLAWAY, L., *Glass Reinforced Plastics in Construction*, 1978, Surrey University Press, Glasgow.

2. LAWRENCE, J. R., *Polyester Resins*, 1960, Van Nostrand Reinhold, New York.
3. MORGAN, P. (ed.), *Glass Reinforced Plastics*, 1955, Iliffe, London.
4. AINSWORTH, L., Fibre-reinforced plastic pipes and applications, *Composites*, **12**, July 1981, 185.
5. AINSWORTH, L., The state of filament winding, *Composites*, **2**, March 1971, 14.
6. BROUTMAN, L. J. and KROCK, R. H., *Modern Composite Materials*, 1967, Addison-Wesley, Reading, Mass.
7. VINSON, J. R. and CHOU, T. W., *Composite Materials and their Use in Structures*, 1975, Applied Science Publishers, London.
8. BENJAMIN, B. S., *Structural Design with Plastics*, 2nd edn., 1982, Van Nostrand Reinhold, New York.
9. GARG, A. C. and TROTMAN, C. K., Influence of water on fracture behaviour of random fibre glass composites, *Engineering Fracture Mechanics*, **13**, 1980, 357.
10. PRITCHARD, G., Environmental degradation of hybrid composite materials, in *Fibre Composite Hybrid Materials*, ed. N. L. Hancox, 1981, Applied Science Publishers, London.
11. ISHAI, O. and MAZOR, A., The effect of environmental-loading history on the transverse strength of GRP laminate, *J. Composite Materials*, **9**, Oct. 1975, 370.

Part B

THE STRUCTURAL PROPERTIES OF GRP

CHAPTER 4

The Theoretical Macromechanical Properties of Composite Laminae

4.1 INTRODUCTION

It has been seen that glass fibre-reinforced materials predominantly take the form of laminates, in which several layers of glass fibres are used to reinforce the resin matrix. Although this form of material is not new, and has been used for many years in timber design, both in the form of laminated timber and in plywood, in glass fibre-reinforced composites the laminations are not as well defined as in the analogous materials, due mainly to the extremely flexible nature and non-uniform length and distribution of the fibres which produce random variations of geometry and strength in the individual layers; hence in any theoretical treatment a number of idealisations and simplifications must be introduced to make any analyses tractable. Thus a theoretical approach to the quantitative determination of the properties of the material must of necessity be tempered by experimental investigations.

In order to develop the analytical treatment of fibre-reinforced composites, it is necessary firstly to examine the behaviour of a single composite lamina, that is a layer of resin reinforced with a single layer of glass fibre, as an understanding of the properties of this single basic entity is a prerequisite to the logical development of the prediction of laminate properties. Such a lamina is of course inherently heterogeneous and its properties along any direction will vary from point to point. However, the heterogeneity and variation in properties are apparent only on the microscopic scale commensurate with the diameter of individual fibres and the distance between them. On a larger scale the lamina may be considered homogeneous and its properties uniform in any given direction, these properties of course being functions of the properties and geometry of the separate components. The treatment of such an assumed homogeneous, though not necessarily isotropic, lamina is known as a *macromechanical* analysis, and it is with the behaviour of such a lamina that this chapter is concerned. The examination of the lamina from the smaller, microscopic viewpoint can yield valuable information regarding the actual strength and stiffness of the material and this will be considered in Chapter 5.

4.2 THE ASSUMPTIONS AND IDEALISATIONS MADE IN THE THEORETICAL TREATMENT OF COMPOSITE LAMINAE

To begin the analytical development of the behaviour of a lamina, it is first necessary to state the idealisations on which such a treatment is based.

Since the properties of the composite lamina depend upon the properties of its components, the nature of the glass fibres and the resin must be defined. It has been mentioned previously that the glass fibres may be considered linear elastic to failure while the resin, although exhibiting reasonably linear behaviour in the working range of stress, becomes nonlinear in the region of the ultimate load. However, since the ultimate strain of the brittle glass fibres is less than that of the more ductile resin, when the lamina is subjected to load in the direction of the fibre reinforcement failure of the glass fibres will occur when the stress in the resin is less than ultimate and hence, prior to fibre failure, the stress in the resin may be reasonably considered low enough to justify the assumption of linear elasticity. Furthermore both constituent materials are considered to be homogeneous and isotropic and also it is assumed that no voids exist and that perfect bonding between fibre and resin occurs. Thus the resulting composite lamina may also be assumed to behave in a linear elastic manner, possibly to incipient failure, and to be, in a macroscopic sense, homogeneous.

The effect of creep in the resin on the stress distribution within the composite lamina and also on its stiffness characteristics may be reduced substantially by ensuring a high proportion of glass fibres in the directions of stress, since as the glass content increases so will the proportion of the applied forces carried by the fibres. The simultaneous reduction of the forces carried by the resin implies that if creep occurs in the matrix, the load thus transferred to the fibres is a relatively small proportion of that already being sustained. On the other hand if the glass fibre content is low then the effect of creep in the resin may be considerable. However, a low fibre content should imply that the applied forces are small enough to ensure little stress-induced creep. The effect of creep is discussed in more detail in Chapter 8 and in the chapters comprising Part C.

Due to the various forms that the fibre reinforcement can take, the resulting lamina, although considered homogeneous, will not necessarily be isotropic, but may display orthotropy to a greater or lesser degree depending upon the nature of the reinforcing medium. Hence the property of isotropy, so common in conventional materials, cannot necessarily be assumed. The effect that this departure has on the stiffness and strength characteristics of a composite lamina is quite profound, as will be seen in the following development.

4.3 THE STRESS–STRAIN RELATIONSHIPS FOR COMPOSITE LAMINAE

In the following development of the relationships between the stresses, or force intensities, and the corresponding deformations in single laminae of composite materials, linear elasticity, equal in tension and compression, will be assumed, as will macroscopic homogeneity. In addition, the deformations will be assumed to be so small that they do not appreciably alter the geometry of the lamina.

These stress–strain relationships are obtained from the consideration of the in-plane static equilibrium and deformation characteristics of an infinitesimally small element at a point in the lamina. Such a consideration is the basis for the development of the theoretical treatment of laminates subjected to both in-plane and lateral forces, as will be seen in Chapter 6.

4.3.1 Isotropic Lamina

A lamina reinforced with a layer of fibres in the form of chopped strand mat may, because of the random orientation of the fibres, reasonably be considered to have, on a macroscopic scale, the same material properties in all directions in the plane of the lamina, and may hence be considered *isotropic*.

Consider an infinitesimally small rectangular element at a point in such a lamina subjected to a state of plane stress, i.e. there is no stress applied normal to the lamina, as shown in Fig. 4.1.

Under any general state of loading the element may be subjected to normal stresses σ_x and σ_y and also to shearing stresses τ_{xy} and τ_{yx}. These

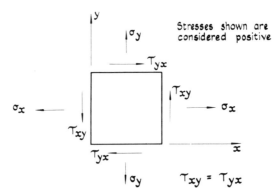

FIG. 4.1. Element subjected to plane stress.

stresses will in general vary from point to point in the lamina, but because the dimensions of the element tend to zero, any stress variations thoughout the element also tend to zero and each stress thus tends to a uniform and constant value. The shearing stresses τ_{xy} and τ_{yx} are necessarily equal since rotation of the element does not occur.[1]

Now because the deformations are considered very small and since linear elasticity in the material is assumed, the relationships between the stresses and strains may be obtained by considering the three stresses applied to the element separately (Fig. 4.2) and combining the results by simple superposition.

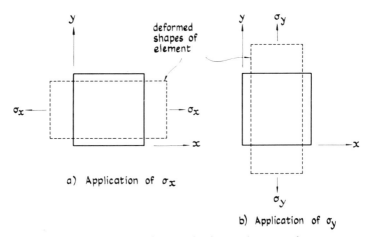

FIG. 4.2. Deformation of isotropic element by normal stresses.

4.3.1.1 Effect of Normal Stresses

On application of a normal tensile stress, σ_x, alone (Fig. 4.2a), the element extends in the x-direction, the ratio of this extension to the original x-directed length being defined as the normal tensile strain in the x-direction. The modulus of elasticity or Young's modulus, E, is then defined as the ratio of the normal stress to this normal strain. Simultaneously the element contracts in the y-direction, thus producing a normal compressive strain in this orthogonal direction. The positive value of the ratio of this latter normal strain to the normal strain in the direction of stress is the elastic material property Poisson's ratio, v. Naturally, if σ_x were compressive then a compressive strain would be experienced in the x-direction with a corresponding extension in the y-direction. Thus the normal stress, σ_x, when applied alone will produce the following normal strains:

(i) in the x-direction, σ_x/E
(ii) in the y-direction, due to Poisson's ratio, $-v\sigma_x/E$

In a similar manner, the application of a normal stress σ_y (Fig. 4.2b) will produce the following two normal strains:

(i) in the y-direction, σ_y/E
(ii) in the x-direction, due to Poisson's ratio, $-v\sigma_y/E$

Hence the total normal strains produced by applying σ_x and σ_y together are obtained by adding together these separate results giving:

$$\text{Total normal strain in } x\text{-direction} = \varepsilon_x = \frac{\sigma_x}{E} - \frac{v\sigma_y}{E}$$

$$= \frac{1}{E}(\sigma_x - v\sigma_y) \qquad (4.1)$$

$$\text{Total normal strain in } y\text{-direction} = \varepsilon_y = \frac{\sigma_y}{E} - \frac{v\sigma_x}{E}$$

$$= \frac{1}{E}(\sigma_y - v\sigma_x) \qquad (4.2)$$

These equations may now be rearranged to express the stresses in terms of the strains, producing the relationships

$$\sigma_x = \frac{E}{1-v^2}(\varepsilon_x + v\varepsilon_y) \tag{4.3}$$

$$\sigma_y = \frac{E}{1-v^2}(\varepsilon_y + v\varepsilon_x) \tag{4.4}$$

At this stage it is convenient to rewrite these equations in the more compact and more visually satisfying matrix form.[2-4] Expressing eqns. (4.1) and (4.2) in this form gives

$$\begin{bmatrix} \varepsilon_x \\ \varepsilon_y \end{bmatrix} = \begin{bmatrix} \dfrac{1}{E} & -\dfrac{v}{E} \\ -\dfrac{v}{E} & \dfrac{1}{E} \end{bmatrix} \begin{bmatrix} \sigma_x \\ \sigma_y \end{bmatrix} \tag{4.5}$$

while eqns. (4.3) and (4.4) can be written as

$$\begin{bmatrix} \sigma_x \\ \sigma_y \end{bmatrix} = \begin{bmatrix} \dfrac{E}{1-v^2} & \dfrac{vE}{1-v^2} \\ \dfrac{vE}{1-v^2} & \dfrac{E}{1-v^2} \end{bmatrix} \begin{bmatrix} \varepsilon_x \\ \varepsilon_y \end{bmatrix} \tag{4.6}$$

Further compaction can be obtained by using only single symbols to represent each matrix. Thus eqns. (4.5) and (4.6) may be written respectively as

$$[\varepsilon] = [S][\sigma] \tag{4.7}$$

and

$$[\sigma] = [Q][\varepsilon] \tag{4.8}$$

In these relationships $[\varepsilon]$ and $[\sigma]$ represent the list of the individual strains and stresses respectively and are known as the strain and stress vectors, while the two square matrices $[S]$ and $[Q]$ describe the relationships between $[\varepsilon]$ and $[\sigma]$. These matrices $[S]$ and $[Q]$ are known as inversions of each other, $[S]$ being termed the *compliance* matrix and $[Q]$ the *material* matrix. It can be noted that the individual components of the [S] and [Q] matrices form a symmetrical pattern about the leading diagonal. This matrix property is a direct consequence of Maxwell's reciprocal theorem.[5-7]

4.3.1.2 Effect of Shearing Stresses

The application of the equal shearing stresses, τ_{xy} and τ_{yx}, on the element causes the element to deform into a parallelogram, the final shape being obtained by the superposition of the effects of the vertical and horizontal shearing stresses, as depicted in Fig. 4.3.

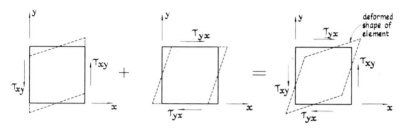

FIG. 4.3. Deformation of isotropic element by shearing stresses.

In order that there be formal similarity between the strains produced by the normal stresses and those produced by the shearing stresses, it should be possible to define shearing strain as a ratio of a deformation to an original length. In the definition of normal strains, the deformation occurs in the same direction as the original length, whereas in the case of distortion by shearing stress, the deformation occurs in a direction normal to an original length, as is clearly seen in the two component diagrams of Fig. 4.3. Thus considering the resultant distortion of the element due to shearing stress as shown in Fig. 4.4, as the deformations

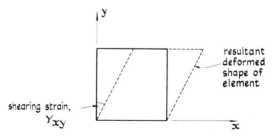

FIG. 4.4. Resultant distortion of isotropic element by shearing stresses and definition of shearing strain.

are small and tend to zero, the required ratio defining the shearing strain tends to the angle γ_{xy} through which two parallel sides of the element rotate relative to the orthogonal sides.

A relationship between the shearing stress and the corresponding shearing strain may thus be defined as

$$\tau_{xy} = G\gamma_{xy} \tag{4.9}$$

where G is known as the *modulus of rigidity*.

This relationship may now be appended to the normal stress–strain equations, and hence the matrix equations (4.5) and (4.6) can be extended to give

$$\begin{bmatrix} \varepsilon_x \\ \varepsilon_y \\ \gamma_{xy} \end{bmatrix} = \begin{bmatrix} \dfrac{1}{E} & -\dfrac{v}{E} & \\ -\dfrac{v}{E} & \dfrac{1}{E} & \\ & & \dfrac{1}{G} \end{bmatrix} \begin{bmatrix} \sigma_x \\ \sigma_y \\ \tau_{xy} \end{bmatrix} \tag{4.10}$$

and

$$\begin{bmatrix} \sigma_x \\ \sigma_y \\ \tau_{xy} \end{bmatrix} = \begin{bmatrix} \dfrac{E}{1-v^2} & \dfrac{vE}{1-v^2} & \\ \dfrac{vE}{1-v^2} & \dfrac{E}{1-v^2} & \\ & & G \end{bmatrix} \begin{bmatrix} \varepsilon_x \\ \varepsilon_y \\ \gamma_{xy} \end{bmatrix} \tag{4.11}$$

These can still be written in the compact forms given in eqns. (4.7) and (4.8), the vectors $[\varepsilon]$ and $[\sigma]$ now comprising three elements, and $[S]$ and $[Q]$ being 3×3 square matrices, symmetrical about the leading diagonal.

Although in general the modulus of rigidity, G, is independent of Young's modulus, E, and Poisson's ratio, v, in the case of an isotropic material G may be expressed as a function of the other two elastic material properties. The relationship between these properties may be deduced from a consideration of a square element, ABCD, subjected to equal tensile and compressive stresses of magnitude σ in the two orthogonal directions x and y and containing within itself a square abcd rotated at $45°$ to the x, y coordinates, as depicted in Fig. 4.5a.

On isolating the triangle Aab and considering equilibrium of this

The theoretical macromechanical properties of composite laminae

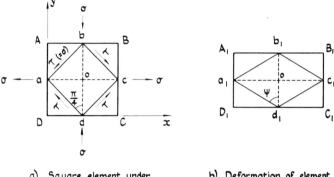

a) Square element under normal stress system

b) Deformation of element

FIG. 4.5. Stress configuration used for deduction of relationship between E, γ and G in isotropic material.

element, it can be deduced that on the face ab there exists only a shearing stress, τ, equal to σ as shown.[1] Similar shearing stresses occur on the remaining faces of the square abcd, and thus the square element abcd is in a state of pure shear.

The distortion of abcd due to shear can now be related to the deformation of ABCD due to the normal stresses. From Fig. 4.5b the shearing strain γ in abcd is equal to $2(\psi - \pi/4)$, where

$$\tan \psi = \frac{Oa_1}{Od_1} = \frac{a_1 c_1}{b_1 d_1}.$$

This ratio of the lengths of the deformed sides can be obtained by considering the effect of the two normal stresses.

Using eqns. (4.1) and (4.2), it can be seen that on considering ABCD,

$$\varepsilon_x = \frac{1}{E}(\sigma + \nu\sigma) = \frac{\sigma}{E}(1+\nu)$$

and

$$\varepsilon_y = \frac{1}{E}(-\sigma - \nu\sigma) = -\frac{\sigma}{E}(1+\nu)$$

That is,

$$\varepsilon_y = -\varepsilon_x$$

Thus

$$\frac{a_1 c_1}{b_1 d_1} = \frac{1+\varepsilon_x}{1+\varepsilon_y} = \frac{1+\varepsilon_x}{1-\varepsilon_x}$$

Now

$$\tan\left(\psi - \frac{\pi}{4}\right) = \frac{\tan\psi - \tan\frac{\pi}{4}}{1+\tan\psi\,\tan\frac{\pi}{4}} = \frac{\tan\psi - 1}{1+\tan\psi}$$

Hence, since

$$\tan\psi = \frac{Oa_1}{Od_1} = \frac{a_1 c_1}{b_1 d_1}$$

$$\tan\left(\psi - \frac{\pi}{4}\right) = \left(\frac{1+\varepsilon_x}{1-\varepsilon_x} - 1\right) \bigg/ \left(1 + \frac{1+\varepsilon_x}{1-\varepsilon_x}\right)$$

$$= \varepsilon_x = \frac{\sigma}{E}(1-v)$$

Now since the shearing strain is small and tends to zero, $\tan(\psi - \pi/4)$ tends to $(\psi - \pi/4)$. Thus the shearing strain is given by

$$\gamma = 2\left(\psi - \frac{\pi}{4}\right) = \frac{2\sigma}{E}(1+v)$$

Hence

$$\frac{\sigma}{\gamma} = \frac{\tau}{\gamma} = G = \frac{E}{2(1+v)} \tag{4.12}$$

This relationship implies that a lamina of an isotropic material can be completely defined by the two elastic material properties, namely Young's modulus, E, and Poisson's ratio, v.

4.3.1.3 Range of Poisson's Ratio and the Extension to Three Dimensions

At this stage it may be noted that a lower limit on the value of Poisson's ratio for an isotropic material is implied by virtue of eqn. (4.12). Since the total work done by the stresses must be positive, a positive shearing stress produces a positive shearing strain, and hence the value of G must be positive, implying that

$$v > -1 \tag{4.13}$$

Although mathematically possible, negative values of Poisson's ratio are not encountered in practice and therefore Poisson's ratio takes a positive value.

However, since a lower limit is seen to exist, the question arises as to whether there is also an upper limit to this property. That such a limit does in fact exist can be deduced from an examination of the behaviour of a three-dimensional block of material as shown in Fig. 4.6a subjected

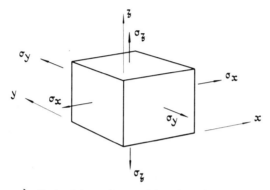

a) Block of isotropic material under normal stresses

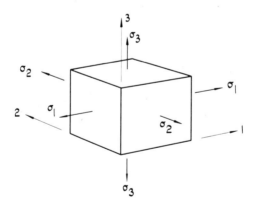

b) Block of orthotropic material under normal stresses applied in principal material directions

Stresses shown are considered positive

FIG. 4.6. Material blocks under normal stress systems.

in turn to normal stresses in one, two, and three orthogonal directions, x, y and z.

Firstly if the block were subjected to a positive normal stress, σ_x, only, then since the total work done by the stress must be positive, the longitudinal strain is positive and thus Young's modulus, E, must have a positive value. In this examination no information is available with regard to the limitations on Poisson's ratio.

However, if secondly the block were acted upon by σ_x and σ_y together, then the stress–strain relationships in the x–y plane are given by eqns. (4.5) and (4.6), and from these equations limitations on the values that Poisson's ratio may take can be deduced. Let the block be restrained from movement in the y-direction, for example by containing it between two immovable walls. Thus ε_y would be eliminated, and the only work done would be that by σ_x in producing the strain ε_x. Hence, since this work must be positive, it follows that the value of $E/(1-v^2)$ must be positive and that therefore the condition

$$1 - v^2 > 0 \qquad (4.14)$$

must be satisfied.

The same result may of course be attained by considering ε_x to be zero and applying σ_y only.

Also it may be noted that the same condition can be obtained from the necessary requirement that for stable equilibrium the determinants of $[S]$ and $[Q]$ in eqns. (4.5) and (4.6) must be positive.[6]

Thus eqn. (4.14) implies that v must lie between -1 and $+1$, the lower limit of this range being that already deduced from eqn. (4.12).

However, the upper limit may still be too high since it has been reduced from the limitless value obtainable from the application of σ_x alone to the value of unity by the application of only σ_x and σ_y together; hence to achieve a definitive upper limit the further case of the block subjected to the three normal stresses σ_x, σ_y and σ_z must be considered. The normal stress–strain equations in three dimensions can be deduced by simply extending the two-dimensional derivations.

Let the block again be subjected to σ_x alone. Due to this stress the longitudinal strain in the x-direction, σ_x/E, is accompanied by lateral strains of $-v\sigma_x/E$ in both the orthogonal directions y and z. Similarly normal stresses σ_y and σ_z will produce corresponding lateral strains in the remaining two orthogonal directions. On combining the separate strains in the three directions it follows that the total strains in the x- and

y-directions are direct extensions of eqns. (4.1) and (4.2), giving

$$\varepsilon_x = \frac{\sigma_x}{E} - \frac{v\sigma_y}{E} - \frac{v\sigma_z}{E}$$

$$= \frac{1}{E}(\sigma_x - v\sigma_y - v\sigma_z) \qquad (4.15)$$

and

$$\varepsilon_y = \frac{\sigma_y}{E} - \frac{v\sigma_z}{E} - \frac{v\sigma_x}{E}$$

$$= \frac{1}{E}(\sigma_y - v\sigma_z - v\sigma_x) \qquad (4.16)$$

and that the total strain in the z-direction is given similarly as

$$\varepsilon_z = \frac{\sigma_z}{E} - \frac{v\sigma_x}{E} - \frac{v\sigma_y}{E}$$

$$= \frac{1}{E}(\sigma_z - v\sigma_x - v\sigma_y) \qquad (4.17)$$

Thus writing these results in matrix form produces

$$\begin{bmatrix} \varepsilon_x \\ \varepsilon_y \\ \varepsilon_z \end{bmatrix} = \begin{bmatrix} \frac{1}{E} & -\frac{v}{E} & -\frac{v}{E} \\ -\frac{v}{E} & \frac{1}{E} & -\frac{v}{E} \\ -\frac{v}{E} & -\frac{v}{E} & \frac{1}{E} \end{bmatrix} \begin{bmatrix} \sigma_x \\ \sigma_y \\ \sigma_z \end{bmatrix} \qquad (4.18)$$

or again more shortly

$$[\varepsilon] = [S][\sigma]$$

On rearranging eqns. (4.18) so that the stresses are given in terms of the strains, the following result is obtained:—

$$\begin{bmatrix} \sigma_x \\ \sigma_y \\ \sigma_z \end{bmatrix} = \frac{E}{(1+v)(1-2v)} \begin{bmatrix} (1-v) & v & v \\ v & (1-v) & v \\ v & v & (1-v) \end{bmatrix} \begin{bmatrix} \varepsilon_x \\ \varepsilon_y \\ \varepsilon_z \end{bmatrix} \qquad (4.19)$$

or in the shorter notation

$$[\sigma] = [Q][\varepsilon]$$

Now, again by allowing only one strain to be non-zero, it follows similarly that

$$\frac{(1-v)}{(1+v)(1-2v)} > 0 \tag{4.20}$$

Since $(1-v)$ must be positive by virtue of eqn. (4.14), it follows that $(1-2v)$ must also be positive and that hence

$$v < \tfrac{1}{2}. \tag{4.21}$$

It can be noted that the lower limit of -1 is also implied in eqn (4.20). Hence Poisson's ratio must lie in the range

$$-1 < v < \tfrac{1}{2}$$

The upper limit may also be obtained by again noting that the determinants of $[S]$ and $[Q]$ in eqns. (4.18) and (4.19) must be positive, conditions that yield the equation

$$1 - 3v^2 - 2v^3 > 0$$

or $\tag{4.22}$

$$(1-2v)(1+v)^2 > 0$$

Indeed in practice these limitations are found to apply to materials that can be reasonably considered isotropic, Poisson's ratio lying within a range between 0 and $\tfrac{1}{2}$.

Although for the purpose of examining the limitations inherent in the value of Poisson's ratio only the normal stress–strain relationships in three dimensions are required, to obtain a complete description of the elastic behaviour of a three-dimensional body the additional shearing stress–strain relationships must be supplied, or in other words eqns. (4.10) and (4.11) must be extended to three dimensions.

The shearing stresses that can act on the faces of a rectilinear block are shown in Fig. 4.7a, and thus, by incorporating these stresses and corresponding strains, eqns (4.18) and (4.19) may be extended to

$$\begin{bmatrix} \varepsilon_x \\ \varepsilon_y \\ \varepsilon_z \\ \gamma_{xy} \\ \gamma_{yz} \\ \gamma_{zx} \end{bmatrix} = \begin{bmatrix} \dfrac{1}{E} & -\dfrac{v}{E} & -\dfrac{v}{E} & & & \\ -\dfrac{v}{E} & \dfrac{1}{E} & -\dfrac{v}{E} & & & \\ -\dfrac{v}{E} & -\dfrac{v}{E} & \dfrac{1}{E} & & & \\ & & & \dfrac{2(1+v)}{E} & & \\ & & & & \dfrac{2(1+v)}{E} & \\ & & & & & \dfrac{2(1+v)}{E} \end{bmatrix} \begin{bmatrix} \sigma_x \\ \sigma_y \\ \sigma_z \\ \tau_{xy} \\ \tau_{yz} \\ \tau_{zx} \end{bmatrix} \quad (4.23)$$

and

$$\begin{bmatrix} \sigma_x \\ \sigma_y \\ \sigma_z \\ \tau_{xy} \\ \tau_{yz} \\ \tau_{zx} \end{bmatrix} = \dfrac{E}{(1+v)(1-2v)} \begin{bmatrix} (1-v) & v & v & & & \\ v & (1-v) & v & & & \\ v & v & (1-v) & & & \\ & & & \dfrac{1-2v}{2} & & \\ & & & & \dfrac{1-2v}{2} & \\ & & & & & \dfrac{1-2v}{2} \end{bmatrix} \begin{bmatrix} \varepsilon_x \\ \varepsilon_y \\ \varepsilon_z \\ \gamma_{xy} \\ \gamma_{yz} \\ \gamma_{zx} \end{bmatrix}$$

(4.24)

G having been written as $E/2(1+v)$.

4.3.2 Orthotropic Lamina — Principal Orientation

In considering laminae reinforced with rovings or fabrics, it is immediately noticed that, unlike those in chopped strand mat, the fibres are aligned in definite directions and not orientated in a random manner.

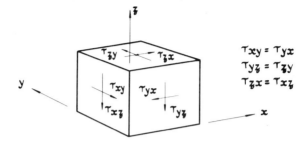

a) Block of isotropic material under shearing stresses

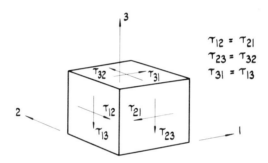

b) Block of orthotropic material under shearing stresses applied in principal material directions

Stresses shown are considered positive

FIG. 4.7. Material blocks under shearing stress systems.

Composites incorporating such forms of reinforcement exhibit high strengths and stiffnesses in the fibre directions, these properties varying with the volume of fibres used. In the case of unidirectional roving reinforcement, for instance, very much higher strengths and stiffnesses will be attained in the direction of the fibres than will be attained normal to the fibre direction, while in woven fabric reinforcement strength and stiffness in the two orthogonal directions will vary according to the volume of glass present in these directions.

Thus it is apparent that such composite laminae will exhibit different values of Young's modulus, Poisson's ratio and modulus of rigidity depending upon the direction in which the stresses and strains are applied.

In composite laminae reinforced with roving or fabric reinforcement it is noticed that the material exhibits symmetry about two axes at right angles to each other (Fig. 4.8). Such a material is known as *orthotropic*, and the properties of such a material are far more complex than those of isotropic materials. In order to develop the elastic material constants of an orthotropic lamina the normal and shearing stresses acting on a rectangular element will again be considered, the effect of each deduced separately and finally the results combined.

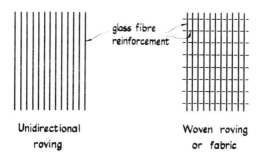

FIG. 4.8. Reinforcement forms in orthotropic laminae.

4.3.2.1 Effect of Normal Stresses

Consider firstly a rectangular element of an orthotropic material subjected to normal stresses in the two principal material directions 1 and 2, i.e. in the directions of the two orthogonal axes about which there is symmetry of the material. Figs (4.9a) and (4.9b) depict a roving-reinforced composite lamina subjected to the separate effects of such stresses. A woven roving fabric reinforced material would of course contain fibres running vertically in addition to those shown running horizontally.

The effect of σ_1 alone produces a strain in the 1-direction equal to

$$\frac{\sigma_1}{E_1}$$

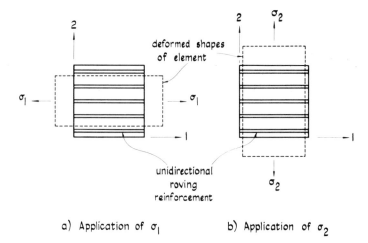

Fig. 4.9. Deformation of orthotropic element by normal stresses applied in principal material directions.

where E_1 is Young's modulus for the material in the 1-direction, and a strain in the 2-direction of

$$-v_{12}\frac{\sigma_1}{E_1}$$

v_{12} being Poisson's ratio for the material when subjected to stress in the 1-direction, the lateral strain taking place in the 2-direction.

Similarly application of σ_2 alone creates a strain in the 2-direction of

$$\frac{\sigma_2}{E_2}$$

and a strain in the 1-direction of

$$-v_{21}\frac{\sigma_2}{E_2}$$

where E_2, v_{21} are respectively Young's modulus for the material in the 2-direction and Poisson's ratio for the material when stressed in the 2-direction, the lateral strain taking place in the 1-direction.

The total normal strains in the 1- and 2-directions, obtained by adding together the respective components, are thus given by:

Total normal strain in 1-direction $= \varepsilon_1 = \dfrac{\sigma_1}{E_1} - v_{21}\dfrac{\sigma_2}{E_2}$ (4.25)

Total normal strain in 2-direction $= \varepsilon_2 = \dfrac{\sigma_2}{E_2} - v_{12}\dfrac{\sigma_1}{E_1}$ (4.26)

or in matrix form

$$\begin{bmatrix} \varepsilon_1 \\ \varepsilon_2 \end{bmatrix} = \begin{bmatrix} \dfrac{1}{E_1} & \dfrac{-v_{21}}{E_2} \\ \dfrac{-v_{12}}{E_1} & \dfrac{1}{E_2} \end{bmatrix} \begin{bmatrix} \sigma_1 \\ \sigma_2 \end{bmatrix}$$ (4.27)

i.e.

$$[\varepsilon] = [S][\sigma]$$

It is now immediately noticed that, since by virtue of Maxwell's reciprocal theorem the compliance matrix $[S]$ must be symmetrical about the leading diagonal,

$$\dfrac{v_{12}}{E_1} = \dfrac{v_{21}}{E_2}$$ (4.28)

This relationship between the elastic material properties implies that the strain in the 2-direction produced by a stress in the 1-direction is the same as the strain in the 1-direction produced by the same stress applied in the 2-direction, although the strains in the directions of stress in the two cases would, in general, be different.

On rearranging eqns. (4.25) and (4.26) (or on inverting the matrix $[S]$ of eqn. (4.27)), the stresses can be expressed in terms of the strains through the material matrix, thus:

$$\begin{bmatrix} \sigma_1 \\ \sigma_2 \end{bmatrix} = \begin{bmatrix} \dfrac{E_1}{1 - v_{12}v_{21}} & \dfrac{v_{21}E_1}{1 - v_{12}v_{21}} \\ \dfrac{v_{12}E_2}{1 - v_{12}v_{21}} & \dfrac{E_2}{1 - v_{12}v_{21}} \end{bmatrix} \begin{bmatrix} \varepsilon_1 \\ \varepsilon_2 \end{bmatrix}$$ (4.29)

i.e.

$$[\sigma] = [Q][\varepsilon]$$

Again the matrix $[Q]$ is symmetrical by virtue of eqn (4.28), and it can also be noticed that if the conditions

$$E_1 = E_2$$

and

$$v_{12} = v_{21}$$

prevail, then eqns. (4.27) and (4.29) degenerate to eqns. (4.5) and (4.6).

4.3.2.2 EFFECT OF SHEARING STRESSES
The effect of shearing stresses applied in the 1,2-directions will, as in the isotropic case, produce distortion of the element from a rectangle to a parallelogram as shown in Fig. 4.10, the ratio of the shearing stress, τ_{12}, to the shearing strain, γ_{12}, being given by G_{12}, the modulus of rigidity of the orthotropic material with respect to the 1- and 2-directions, thus:

$$\tau_{12} = G_{12} \gamma_{12} \tag{4.30}$$

It must be noted here that, unlike the isotropic case, the modulus of rigidity must be defined with respect to the directions in which the stresses and distortions occur, and also that the modulus of rigidity is a separate elastic constant, independent of the values of Young's moduli and Poisson's ratios.

Again the shearing stress–shearing strain relationship can be added to the normal stress–strain equations so that the matrix equations (4.27)

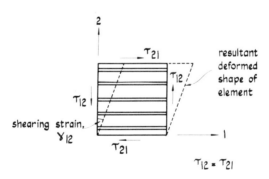

FIG. 4.10. Resultant distortion of orthotropic element by shearing stresses applied in principal material directions.

and (4.29) may be extended to give

$$\begin{bmatrix} \varepsilon_1 \\ \varepsilon_2 \\ \gamma_{12} \end{bmatrix} = \begin{bmatrix} \dfrac{1}{E_1} & -\dfrac{v_{21}}{E_2} & \\ \dfrac{-v_{12}}{E_1} & \dfrac{1}{E_2} & \\ & & \dfrac{1}{G_{12}} \end{bmatrix} \begin{bmatrix} \sigma_1 \\ \sigma_2 \\ \tau_{12} \end{bmatrix} \quad (4.31)$$

and

$$\begin{bmatrix} \sigma_1 \\ \sigma_2 \\ \tau_{12} \end{bmatrix} = \begin{bmatrix} \dfrac{E_1}{1-v_{12}v_{21}} & \dfrac{v_{21}E_1}{1-v_{12}v_{21}} & \\ \dfrac{v_{12}E_2}{1-v_{12}v_{21}} & \dfrac{E_2}{1-v_{12}v_{21}} & \\ & & G_{12} \end{bmatrix} \begin{bmatrix} \varepsilon_1 \\ \varepsilon_2 \\ \gamma_{12} \end{bmatrix} \quad (4.32)$$

Hence a lamina of an orthotropic material is completely defined by the four independent elastic constants E_1, E_2, v_{12} (or v_{21}) and G_{12}.

4.3.2.3 LIMITATIONS ON POISSON'S RATIO VALUES AND EXTENSION TO THREE DIMENSIONS

As in the case of isotropic materials, certain limitations exist in orthotropic materials on the values that Poisson's ratio may take, and these constraints may be deduced in a similar manner to those for istropic materials.

Consider a rectilinear block of orthotropic material whose sides lie along the orthogonal system of axes 1, 2, 3 (Fig. 4.6b). Here the further principal material direction, 3, has been introduced orthogonal to the two already considered. In the three directions 1, 2, and 3 the material has values of Young's modulus E_1, E_2 and E_3 respectively, and with respect to the three planes 1–2, 2–3, 3–1, has values of Poisson's ratio v_{12}, v_{21}, v_{23}, v_{32}, v_{31}, v_{13} and modulus of rigidity G_{12}, G_{23} and G_{31}. The block is now subjected to various combinations of normal stress.

Firstly, if the block is acted upon by σ_1, σ_2, and σ_3 separately then as before it follows that E_1, E_2, and E_3 are positive. Similarly the three values of modulus of rigidity are positive.

Secondly, let the block be subjected to stresses σ_1 and σ_2 together. The relationship between these stresses and the corresponding strains is given by eqns. (4.27) and (4.29), and by the same reasoning as in the isotropic case, it follows from eqn. (4.29) that $E_1/(1-v_{12}v_{21})$ and $E_2/(1-v_{12}v_{21})$ must both be positive, and hence that the condition

$$1 - v_{12}v_{21} > 0 \qquad (4.33)$$

must prevail.

As in the isotropic case, this result can be obtained from the condition that the determinants of the $[S]$ and $[Q]$ matrices in eqns. (4.27) and (4.29) must be positive.

Now using eqn. (4.28) it can be seen that the condition given by eqn. (4.33) may be expressed as

$$1 - v_{12}v_{21} = 1 - v_{12}^2 \frac{E_2}{E_1} = 1 - v_{21}^2 \frac{E_1}{E_2} > 0 \qquad (4.34)$$

i.e.

$$v_{12}^2 < \frac{E_1}{E_2} \qquad (4.35a)$$

and

$$v_{21}^2 < \frac{E_2}{E_1} \qquad (4.35b)$$

Similar results apply for the block subjected to stresses in the 2–3 and 3–1 planes.

Extension of the normal orthotropic stress–strain equations to three dimensions will, as in the isotropic case, produce a further constraint on the values that Poisson's ratio may take. By subjecting the block to all three normal stresses, the total normal strain in each of the three orthogonal planes will be extensions of eqns. (4.25) and (4.26), thus:

$$\varepsilon_1 = \frac{\sigma_1}{E_1} - v_{21}\frac{\sigma_2}{E_2} - v_{31}\frac{\sigma_3}{E_3} \qquad (4.36)$$

$$\varepsilon_2 = \frac{\sigma_2}{E_2} - v_{32}\frac{\sigma_3}{E_3} - v_{12}\frac{\sigma_1}{E_1} \qquad (4.37)$$

$$\varepsilon_3 = \frac{\sigma_3}{E_3} - v_{13}\frac{\sigma_1}{E_1} - v_{23}\frac{\sigma_2}{E_2} \qquad (4.38)$$

and

$$\begin{bmatrix} \sigma_1 \\ \sigma_2 \\ \sigma_3 \\ \tau_{12} \\ \tau_{23} \\ \tau_{31} \end{bmatrix} = \frac{1}{\Delta} \begin{bmatrix} E_1(1-v_{23}v_{32}) & E_1(v_{21}+v_{31}v_{23}) & E_1(v_{31}+v_{21}v_{32}) & & & \\ E_2(v_{12}+v_{13}v_{32}) & E_2(1-v_{13}v_{31}) & E_2(v_{32}+v_{12}v_{31}) & & & \\ E_3(v_{13}+v_{12}v_{23}) & E_3(v_{23}+v_{21}v_{13}) & E_3(1-v_{12}v_{21}) & & & \\ & & & \Delta G_{12} & & \\ & & & & \Delta G_{23} & \\ & & & & & \Delta G_{31} \end{bmatrix} \begin{bmatrix} \varepsilon_1 \\ \varepsilon_2 \\ \varepsilon_3 \\ \gamma_{12} \\ \gamma_{23} \\ \gamma_{31} \end{bmatrix}$$

(4.44)

the directions of the shearing stresses being shown in Fig. 4.7(b).

4.3.3 Orthotropic Lamina — Arbitrary Orientation

It is important to notice that the relationships between the stresses and strains given in eqns. (4.43) and (4.44) are defined relative to the principal directions of the material. These particular orientations of the stresses and corresponding strains mean that, because of material symmetry, the effects of the normal stresses are independent of the effects of shearing stresses, and hence the two effects can be considered separately.

Although many composite laminae are amenable to this form of treatment, there can also arise cases where the stresses and strains required to be computed are not described with respect to the principal material directions 1,2 but are defined relative to another set of axes x, y at some arbitrary orientation, θ, to them. This form of lamina is found for instance in helically-wound cylindrical structures such as those produced by filament winding, and also in laminates composed of a number of laminae at various orientations. An element of such a lamina is shown in Fig. 4.11a.

In order to obtain the relationships between the stresses and strains relative to a set of axes arbitrarily orientated to the axes containing the principal directions of the material, the relationships between the stresses defined with respect to the two sets of axes in a lamina must first be considered, followed by the similar relationship between the corresponding strains. The transformation relationship between the axes is shown in Fig. 4.11b, the 1–2 set being the principal material axes of the lamina and the x–y set being the axes in which the stresses and strains are to be obtained.

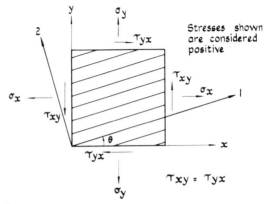

a) Othotropic lamina subjected to stresses defined about an arbitrary coordinate system

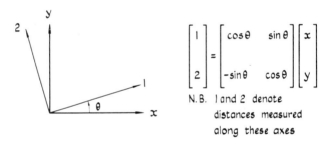

b) Coordinate transformation relationship

FIG. 4.11. Arbitrary orientation of orthotropic lamina.

4.3.3.1 RELATIONSHIPS BETWEEN STRESSES

To find the relationships between the stresses, the equilibrium of two triangular elements ABC of unit thickness subjected to these stresses is considered, as shown in Fig. 4.12.

Referring to Fig. 4.12a, resolution of the forces in the 1-direction gives the equilibrium equation

$$\sigma_1 \cdot BC - \sigma_x AC \cos\theta - \sigma_y AB \sin\theta - \tau_{xy} AC \sin\theta$$
$$- \tau_{yx} AB \cos\theta = 0 \qquad (4.45a)$$

Hence, since $AC = BC \cos\theta$ and $AB = BC \sin\theta$, and since $\tau_{xy} = \tau_{yx}$,

$$\sigma_1 = \sigma_x \cos^2\theta + \sigma_y \sin^2\theta + \tau_{xy} 2\sin\theta\cos\theta \qquad (4.45b)$$

The theoretical macromechanical properties of composite laminae 73

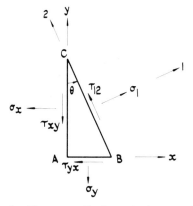

a) Element utilised to obtain σ_1 and τ_{12}

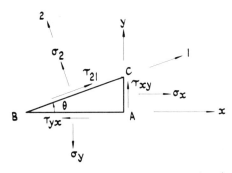

b) Element utilised to obtain σ_2 and τ_{21} ($=\tau_{12}$)

FIG. 4.12. Elements used for determining relationships between stresses in two orthogonal coordinate systems.

Similarly, resolving forces in the 2-direction gives

$$\tau_{12} \cdot BC + \sigma_x AC \sin\theta - \sigma_y AB \cos\theta - \tau_{xy} AC \cos\theta$$
$$+ \tau_{yx} AB \sin\theta = 0 \qquad (4.46a)$$

which yields

$$\tau_{12} = -\sigma_x \sin\theta \cos\theta + \sigma_y \sin\theta \cos\theta$$
$$+ \tau_{xy} (\cos^2\theta - \sin^2\theta) \qquad (4.46b)$$

In a similar manner σ_2 may be obtained by considering the equilibrium of the triangle in Fig. 4.12b in the 2-direction, thus

$$\sigma_2 \cdot BC - \sigma_x AC \sin\theta - \sigma_y AB \cos\theta + \tau_{xy} AC \cos\theta$$
$$+ \tau_{yx} AB \sin\theta = 0 \qquad (4.47a)$$

which, since $AC = BC \sin\theta$ and $AB = BC \cos\theta$, gives

$$\sigma_2 = \sigma_x \sin^2\theta + \sigma_y \cos^2\theta - \tau_{xy} 2\sin\theta\cos\theta \qquad (4.47b)$$

The value of τ_{12} may be similarly re-established by considering equilibrium in the 1-direction.

Expressing eqns. (4.45b), (4.46b), and (4.47b) in matrix form gives

$$\begin{bmatrix} \sigma_1 \\ \sigma_2 \\ \tau_{12} \end{bmatrix} = \begin{bmatrix} \cos^2\theta & \sin^2\theta & 2\sin\theta\cos\theta \\ \sin^2\theta & \cos^2\theta & -2\sin\theta\cos\theta \\ -\sin\theta\cos\theta & \sin\theta\cos\theta & (\cos^2\theta - \sin^2\theta) \end{bmatrix} \begin{bmatrix} \sigma_x \\ \sigma_y \\ \tau_{xy} \end{bmatrix}$$

(4.48)

or more shortly

$$\begin{bmatrix} \sigma_1 \\ \sigma_2 \\ \tau_{12} \end{bmatrix} = [T] \begin{bmatrix} \sigma_x \\ \sigma_y \\ \tau_{xy} \end{bmatrix} \qquad (4.49)$$

These equations may of course be written so that the stresses in the x–y system are expressed in terms of those in the 1–2 coordinates thus:

$$\begin{bmatrix} \sigma_x \\ \sigma_y \\ \tau_{xy} \end{bmatrix} = \begin{bmatrix} \cos^2\theta & \sin^2\theta & -2\sin\theta\cos\theta \\ \sin^2\theta & \cos^2\theta & 2\sin\theta\cos\theta \\ \sin\theta\cos\theta & -\sin\theta\cos\theta & (\cos^2\theta - \sin^2\theta) \end{bmatrix} \begin{bmatrix} \sigma_1 \\ \sigma_2 \\ \tau_{12} \end{bmatrix}$$

(4.50)

$$\text{or} \quad \begin{bmatrix} \sigma_x \\ \sigma_y \\ \tau_{xy} \end{bmatrix} = [T]^{-1} \begin{bmatrix} \sigma_1 \\ \sigma_2 \\ \tau_{12} \end{bmatrix} \qquad (4.51)$$

where $[T]^{-1}$ denotes the inverse of $[T]$.

4.3.3.2 RELATIONSHIPS BETWEEN STRAINS

To determine the relationships between the strains related to the 1–2 axes and those in the x–y system, consider now a rectangular element of a lamina, whose sides lie in the x, y-directions, both before and after deformation induced by plane stress. This is depicted in Fig. 4.13a, where u and v are the small displacements of point A in the x-and y-directions respectively, these displacements in general varying throughout the element due to straining.

Since in general every point in the element may undergo displacement, the deformation of the element may be thought of as being composed of two forms of displacement, one due to normal and shearing strains and one due to the movement of the element as a rigid body. The rigid body

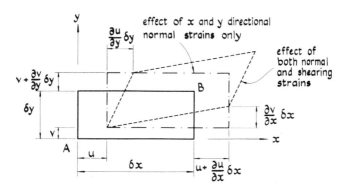

a) General deformation of element

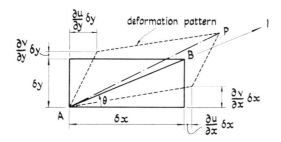

b) Deformation of element after elimination of rigid body displacements

FIG. 4.13. Deformation of rectangular element.

displacement has no effect on the strains and therefore, as far as the calculation of the strains is concerned, may be eliminated, and the position of the deformed element considered to be as shown in Fig. 4.13b. From this diagram it can be noted that the normal strains in the x- and y-directions are given by

$$\varepsilon_x = \frac{\partial u}{\partial x} \delta x / \delta x = \frac{\partial u}{\partial x} \quad (4.52)$$

$$\varepsilon_y = \frac{\partial v}{\partial y} \delta y / \delta y = \frac{\partial v}{\partial y} \quad (4.53)$$

while the shearing strain related to the x- and y-directions can be expressed as

$$\gamma_{xy} = \frac{\partial u}{\partial y} \delta y / \delta y + \frac{\partial v}{\partial x} \delta x / \delta x = \frac{\partial u}{\partial y} + \frac{\partial v}{\partial x} \quad (4.54)$$

Now on examination in Fig. 4.13b of the line AB which makes an angle of θ with the x-direction, and hence from Fig. 4.11 becomes the 1-axis, it can be seen that in moving to P, point B undergoes a displacement which may be expressed in terms of components either in the x- and y-directions or in the 1- and 2-directions.

These components are shown in Fig. 4.14, those in the x- and y-directions being BR and PR, and those in the 1- and 2-directions being BQ and PQ respectively.

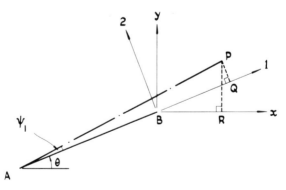

FIG. 4.14. Component displacements of B in x, y- and 1, 2-directions.

Referring again to Fig. 4.13b, the values of the components in the x- and y-directions can be seen to be given by

$$\mathrm{BR} = \frac{\partial u}{\partial x}\delta x + \frac{\partial u}{\partial y}\delta y \tag{4.55}$$

and

$$\mathrm{PR} = \frac{\partial v}{\partial x}\delta x + \frac{\partial v}{\partial y}\delta y \tag{4.56}$$

where partial differential coefficients are used since u and v are in general functions of both x and y.

Now BQ and PQ, the components of the displacement of B in the 1- and 2-directions, can be expressed in terms of those in the x- and y-directions, namely BR and PR, by the standard transformation of axes shown in Fig. 4.11b.

Thus

$$\begin{bmatrix}\mathrm{BQ}\\\mathrm{PQ}\end{bmatrix} = \begin{bmatrix}\cos\theta & \sin\theta\\-\sin\theta & \cos\theta\end{bmatrix}\begin{bmatrix}\mathrm{BR}\\\mathrm{PR}\end{bmatrix} \tag{4.57}$$

which, on using eqns. (4.55) and (4.56), become

$$\begin{bmatrix}\mathrm{BQ}\\\mathrm{PQ}\end{bmatrix} = \begin{bmatrix}\cos\theta & \sin\theta\\-\sin\theta & \cos\theta\end{bmatrix}\begin{bmatrix}\frac{\partial u}{\partial x}\delta x + \frac{\partial u}{\partial y}\delta_y\\\frac{\partial v}{\partial x}\delta x + \frac{\partial v}{\partial y}\delta_y\end{bmatrix} \tag{4.58}$$

These equations may of course be written separately as

$$\mathrm{BQ} = \cos\theta\left(\frac{\partial u}{\partial x}\delta x + \frac{\partial u}{\partial y}\delta y\right) + \sin\theta\left(\frac{\partial v}{\partial x}\delta x + \frac{\partial v}{\partial y}\delta y\right) \tag{4.59}$$

and

$$\mathrm{PQ} = -\sin\theta\left(\frac{\partial u}{\partial x}\delta x + \frac{\partial u}{\partial y}\delta y\right) + \cos\theta\left(\frac{\partial v}{\partial x}\delta x + \frac{\partial v}{\partial y}\delta y\right) \tag{4.60}$$

Now since the displacements tend to zero, the actual deformed length of AB, that is AP, tends to the length AQ, and the length PQ tends to the displacement of B normal to AB. Hence, letting the length of AB be δs, and using eqn. (4.59), the value of the normal strain in the 1-direction may be written as

$$\varepsilon_1 = \frac{BQ}{AB} = \cos\theta \left(\frac{\partial u}{\partial x}\frac{\delta x}{\delta s} + \frac{\partial u}{\partial y}\frac{\delta y}{\delta s} \right) + \sin\theta \left(\frac{\partial v}{\partial x}\frac{\delta x}{\delta s} + \frac{\partial v}{\partial y}\frac{\delta y}{\delta s} \right) \quad (4.61)$$

which, since

$$\frac{\delta x}{\delta s} = \cos\theta \quad (4.62a)$$

and

$$\frac{\delta y}{\delta s} = \sin\theta, \quad (4.62b)$$

may be expressed as

$$\varepsilon_1 = \frac{\partial u}{\partial x}\cos^2\theta + \frac{\partial v}{\partial y}\sin^2\theta + \left(\frac{\partial u}{\partial y} + \frac{\partial v}{\partial x} \right)\sin\theta\cos\theta \quad (4.63)$$

that is, invoking eqns. (4.52), (4.53) and (4.54),

$$\varepsilon_1 = \varepsilon_x \cos^2\theta + \varepsilon_y \sin^2\theta + \frac{\gamma_{xy}}{2} 2\sin\theta\cos\theta \quad (4.64)$$

At this stage it is noticed that the form of eqn. (4.64) is exactly the same as that of eqn. (4.45b) with the normal strains taking the place of the corresponding normal stresses and half the shearing strain replacing the shearing stress. It may therefore be expected that the other two strain equations, namely those for normal strain in the 2-direction and for shearing strain related to the 1- and 2-axes will similarly follow their counterpart stress equations, and this, as will now be shown, is indeed the case.

In the same manner as ε_1 was obtained, an expression for the angle BAP (Figs. 4.13b and 4.14) may be obtained by dividing PQ by AB. Letting angle BAP be denoted by ψ_1 and using eqn. (4.60), together with the relationships given by eqns. (4.62a,b), it follows that

$$\psi_1 = \frac{PQ}{AB} = \frac{\partial v}{\partial x}\cos^2\theta + \left(\frac{\partial v}{\partial y} - \frac{\partial u}{\partial x} \right)\sin\theta\cos\theta - \frac{\partial u}{\partial y}\sin^2\theta \quad (4.65)$$

If the 1- and 2-axes are now rotated anticlockwise through 90° ($\pi/2$ radians) to 1' and 2', expressions for ε_2 and another angle ψ_2 similar to eqns. (4.64) and (4.65) can be written by replacing θ by $(\theta + \pi/2)$. Hence, noting that axes 1' and 2 coincide,

$$\varepsilon_1' = \varepsilon_2 = \varepsilon_x \cos^2\left(\theta + \frac{\pi}{2}\right) + \varepsilon_y \sin^2\left(\theta + \frac{\pi}{2}\right) +$$

$$\frac{\gamma_{xy}}{2} 2 \sin\left(\theta + \frac{\pi}{2}\right) \cos\left(\theta + \frac{\pi}{2}\right) \tag{4.66}$$

which on simplifying becomes

$$\varepsilon_2 = \varepsilon_x \sin^2\theta + \varepsilon_y \cos^2\theta - \frac{\gamma_{xy}}{2} 2 \sin\theta \cos\theta \tag{4.67}$$

Similarly

$$\psi_2 = \frac{\partial v}{\partial x} \sin^2\theta - \left(\frac{\partial v}{\partial y} - \frac{\partial u}{\partial x}\right) \sin\theta \cos\theta - \frac{\partial u}{\partial y} \cos^2\theta \tag{4.68}$$

the nature of this angle being clarified in Fig. 4.15.

The equation for the shearing strain can now be obtained from the two angular expressions ψ_1 and ψ_2 given by eqns. (4.65) and (4.68). From

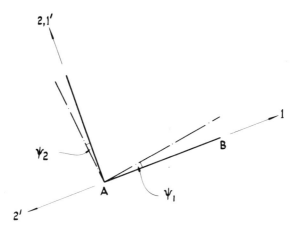

FIG. 4.15. Nature of angles ψ_1 and ψ_2.

Fig. 4.15 it can be seen that the shearing strain related to the 1- and 2-axes is given by

$$\gamma_{12} = \psi_1 - \psi_2 \tag{4.69}$$

that is, the relative rotation of lines in the 1- and 2-directions; therefore, on substituting from eqns. (4.65) and (4.68),

$$\gamma_{12} = \left(\frac{\partial v}{\partial y} - \frac{\partial u}{\partial x}\right) 2 \sin\theta \cos\theta + \frac{\partial v}{\partial x}(\cos^2\theta - \sin^2\theta)$$

$$+ \frac{\partial u}{\partial y}(\cos^2\theta - \sin^2\theta)$$

$$= -\varepsilon_x 2 \sin\theta \cos\theta + \varepsilon_y 2 \sin\theta \cos\theta$$

$$+ \gamma_{xy}(\cos^2\theta - \sin^2\theta) \tag{4.70}$$

On dividing by 2 throughout, the result

$$\frac{\gamma_{12}}{2} = -\varepsilon_x \sin\theta \cos\theta + \varepsilon_y \sin\theta \cos\theta$$

$$+ \frac{\gamma_{xy}}{2}(\cos^2\theta - \sin^2\theta) \tag{4.71}$$

is obtained, which is of the same form as eqn. (4.46b) for the corresponding shearing stress.

Thus expressing the strain relationships given by eqns. (4.64), (4.67) and (4.71) in matrix form gives

$$\begin{bmatrix} \varepsilon_1 \\ \varepsilon_2 \\ \frac{\gamma_{12}}{2} \end{bmatrix} = \begin{bmatrix} \cos^2\theta & \sin^2\theta & 2\sin\theta\cos\theta \\ \sin^2\theta & \cos^2\theta & -2\sin\theta\cos\theta \\ -\sin\theta\cos\theta & \sin\theta\cos\theta & (\cos^2\theta - \sin^2\theta) \end{bmatrix} \begin{bmatrix} \varepsilon_x \\ \varepsilon_y \\ \frac{\gamma_{xy}}{2} \end{bmatrix}$$

$$\tag{4.72}$$

or more briefly

$$\begin{bmatrix} \varepsilon_1 \\ \varepsilon_2 \\ \dfrac{\gamma_{12}}{2} \end{bmatrix} = [T] \begin{bmatrix} \varepsilon_x \\ \varepsilon_y \\ \dfrac{\gamma_{xy}}{2} \end{bmatrix} \qquad (4.73)$$

which upon inversion becomes

$$\begin{bmatrix} \varepsilon_x \\ \varepsilon_y \\ \dfrac{\gamma_{xy}}{2} \end{bmatrix} = [T]^{-1} \begin{bmatrix} \varepsilon_1 \\ \varepsilon_2 \\ \dfrac{\gamma_{12}}{2} \end{bmatrix} \qquad (4.74)$$

In other words, the stress and strain transformations are identical if in the strain equations the normal strains and half the shearing strain are used in place of the normal and shearing stresses.

4.3.3.3 STRESS–STRAIN RELATIONSHIPS

The derivation of the matrix $[T]$ now enables the relationships between the stresses and strains defined in the arbitrary x–y system to be obtained in terms of the material properties in the 1–2 coordinates in a very straightforward manner.

Equations (4.31) and (4.32) may be written respectively as

$$\begin{bmatrix} \varepsilon_1 \\ \varepsilon_2 \\ \gamma_{12} \end{bmatrix} = [S] \begin{bmatrix} \sigma_1 \\ \sigma_2 \\ \tau_{12} \end{bmatrix} \qquad (4.75)$$

and

$$\begin{bmatrix} \sigma_1 \\ \sigma_2 \\ \tau_{12} \end{bmatrix} = [Q] \begin{bmatrix} \varepsilon_1 \\ \varepsilon_2 \\ \gamma_{12} \end{bmatrix} \qquad (4.76)$$

these being the stress–strain relationships referred to the principal material directions.

Hence, by using the stress transformation equations (4.49) and (4.51), the stress vectors in eqns. (4.75) and (4.76) can be transformed from the 1–2 system to the x–y system of coordinates, thus:

$$\begin{bmatrix} \varepsilon_1 \\ \varepsilon_2 \\ \gamma_{12} \end{bmatrix} = [S][T] \begin{bmatrix} \sigma_x \\ \sigma_y \\ \tau_{xy} \end{bmatrix} \tag{4.77}$$

and

$$\begin{bmatrix} \sigma_x \\ \sigma_y \\ \tau_{xy} \end{bmatrix} = [T]^{-1}[Q] \begin{bmatrix} \varepsilon_1 \\ \varepsilon_2 \\ \gamma_{12} \end{bmatrix} \tag{4.78}$$

In order similarly to transform the strain vectors, it is convenient first to modify the analogous strain transformation equations (4.73) and (4.74) so that the shearing strain terms revert from their halved to their full values.

This is attained by using the matrix[3]

$$[R] = \begin{bmatrix} 1 & 0 & 0 \\ 0 & 1 & 0 \\ 0 & 0 & 2 \end{bmatrix} \tag{4.79}$$

Hence

$$\begin{bmatrix} \varepsilon_1 \\ \varepsilon_2 \\ \gamma_{12} \end{bmatrix} = [R] \begin{bmatrix} \varepsilon_1 \\ \varepsilon_2 \\ \dfrac{\gamma_{12}}{2} \end{bmatrix} \tag{4.80}$$

and

$$\begin{bmatrix} \varepsilon_x \\ \varepsilon_y \\ \gamma_{xy} \end{bmatrix} = [R] \begin{bmatrix} \varepsilon_x \\ \varepsilon_y \\ \dfrac{\gamma_{xy}}{2} \end{bmatrix} \tag{4.81}$$

The theoretical macromechanical properties of composite laminae 83

while similarly

$$\begin{bmatrix} \varepsilon_1 \\ \varepsilon_2 \\ \dfrac{\gamma_{12}}{2} \end{bmatrix} = [R]^{-1} \begin{bmatrix} \varepsilon_1 \\ \varepsilon_2 \\ \gamma_{12} \end{bmatrix} \quad (4.82)$$

and

$$\begin{bmatrix} \varepsilon_x \\ \varepsilon_y \\ \dfrac{\gamma_{xy}}{2} \end{bmatrix} = [R]^{-1} \begin{bmatrix} \varepsilon_x \\ \varepsilon_y \\ \gamma_{xy} \end{bmatrix} \quad (4.83)$$

where $[R]^{-1}$ is the inverse of $[R]$, i.e.

$$\begin{bmatrix} 1 & 0 & 0 \\ 0 & 1 & 0 \\ 0 & 0 & \tfrac{1}{2} \end{bmatrix}$$

Thus the strain transformation equations (4.73) and (4.74) may be reformed as

$$\begin{bmatrix} \varepsilon_1 \\ \varepsilon_2 \\ \gamma_{12} \end{bmatrix} = [R][T][R]^{-1} \begin{bmatrix} \varepsilon_x \\ \varepsilon_y \\ \gamma_{xy} \end{bmatrix} \quad (4.84)$$

$$\begin{bmatrix} \varepsilon_x \\ \varepsilon_y \\ \gamma_{xy} \end{bmatrix} = [R][T]^{-1}[R]^{-1} \begin{bmatrix} \varepsilon_1 \\ \varepsilon_2 \\ \gamma_{12} \end{bmatrix} \quad (4.85)$$

On carrying out these simple triple matrix products it is found that

$$[R][T][R]^{-1} = \begin{bmatrix} \cos^2\theta & \sin^2\theta & \sin\theta\cos\theta \\ \sin^2\theta & \cos^2\theta & -\sin\theta\cos\theta \\ -2\sin\theta\cos\theta & 2\sin\theta\cos\theta & (\cos^2\theta - \sin^2\theta) \end{bmatrix}$$

(4.86)

and

$$[R][T]^{-1}[R]^{-1} = \begin{bmatrix} \cos^2\theta & \sin^2\theta & -\sin\theta\cos\theta \\ \sin^2\theta & \cos^2\theta & \sin\theta\cos\theta \\ 2\sin\theta\cos\theta & -2\sin\theta\cos\theta & (\cos^2\theta - \sin^2\theta) \end{bmatrix}$$
(4.87)

Now if these matrices are compared with $[T]$ and $[T]^{-1}$ in eqns. (4.48), (4.72) and (4.50), it is noticed that

$$[R][T][R]^{-1} = \{[T]^{-1}\}^T \tag{4.88}$$

and

$$[R][T]^{-1}[R]^{-1} = [T]^T \tag{4.89}$$

where the superscript T denotes the transpose of the matrices, that is the matrices with the rows and columns interchanged.

Using these relationships the strain transformation equations (4.84) and (4.85) can be expressed very compactly as

$$\begin{bmatrix} \varepsilon_1 \\ \varepsilon_2 \\ \gamma_{12} \end{bmatrix} = \{[T]^{-1}\}^T \begin{bmatrix} \varepsilon_x \\ \varepsilon_y \\ \gamma_{xy} \end{bmatrix} \tag{4.90}$$

and

$$\begin{bmatrix} \varepsilon_x \\ \varepsilon_y \\ \gamma_{xy} \end{bmatrix} = [T]^T \begin{bmatrix} \varepsilon_1 \\ \varepsilon_2 \\ \gamma_{12} \end{bmatrix} \tag{4.91}$$

and hence finally, upon substituting eqns. (4.90) and (4.91), eqns. (4.77) and (4.78) can be further developed to yield

$$\begin{bmatrix} \varepsilon_x \\ \varepsilon_y \\ \gamma_{xy} \end{bmatrix} = [T]^T[S][T] \begin{bmatrix} \sigma_x \\ \sigma_y \\ \tau_{xy} \end{bmatrix} = [\bar{S}] \begin{bmatrix} \sigma_x \\ \sigma_y \\ \tau_{xy} \end{bmatrix} \tag{4.92}$$

and

$$\begin{bmatrix} \sigma_x \\ \sigma_y \\ \tau_{xy} \end{bmatrix} = [T]^{-1}[Q]\{[T]^{-1}\}^T \begin{bmatrix} \varepsilon_x \\ \varepsilon_y \\ \gamma_{xy} \end{bmatrix} = [\bar{Q}] \begin{bmatrix} \varepsilon_x \\ \varepsilon_y \\ \gamma_{xy} \end{bmatrix} \quad (4.93)$$

On performing the triple matrix product $[T]^T[S][T]$ in eqn. (4.92) the relationships between the stresses and strains in the arbitrary x–y coordinate system become

$$\begin{bmatrix} \varepsilon_x \\ \varepsilon_y \\ \gamma_{xy} \end{bmatrix} = \begin{bmatrix} \bar{S}_{11} & \bar{S}_{12} & \bar{S}_{13} \\ \bar{S}_{21} & \bar{S}_{22} & \bar{S}_{23} \\ \bar{S}_{31} & \bar{S}_{32} & \bar{S}_{33} \end{bmatrix} \begin{bmatrix} \sigma_x \\ \sigma_y \\ \tau_{xy} \end{bmatrix} \quad (4.94)$$

where

$$\bar{S}_{11} = S_{11} \cos^4 \theta + S_{22} \sin^4 \theta + (2S_{12} + S_{33}) \sin^2 \theta \cos^2 \theta$$

$$\bar{S}_{12} = \bar{S}_{21} = S_{12} (\sin^4 \theta + \cos^4 \theta) \\ + (S_{11} + S_{22} - S_{33}) \sin^2 \theta \cos^2 \theta$$

$$\bar{S}_{13} = \bar{S}_{31} = (2S_{11} - 2S_{12} - S_{33}) \sin \theta \cos^3 \theta \\ - (2S_{22} - 2S_{12} - S_{33}) \sin^3 \theta \cos \theta$$

$$\bar{S}_{22} = S_{11} \sin^4 \theta + S_{22} \cos^4 \theta + (2S_{12} + S_{33}) \sin^2 \theta \cos^2 \theta$$

$$\bar{S}_{23} = \bar{S}_{32} = (2S_{11} - 2S_{12} - S_{33}) \sin^3 \theta \cos \theta \\ - (2S_{22} - 2S_{12} - S_{33}) \sin \theta \cos^3 \theta$$

$$\bar{S}_{33} = 2(2S_{11} + 2S_{22} - 4S_{12} - S_{33}) \sin^2 \theta \cos^2 \theta \\ + S_{33}(\sin^4 \theta + \cos^4 \theta)$$

and from eqns. (4.31),

$$S_{11} = \frac{1}{E_1}$$

$$S_{22} = \frac{1}{E_2}$$

$$S_{12} = -\frac{v_{21}}{E_2} = -\frac{v_{12}}{E_1}$$

$$S_{33} = \frac{1}{G_{12}}$$

Similarly on calculating $[T]^{-1}[Q]\{[T]^{-1}\}^T$ in eqns. (4.93), the inverse relationship

$$\begin{bmatrix} \sigma_x \\ \sigma_y \\ \tau_{xy} \end{bmatrix} = \begin{bmatrix} \bar{Q}_{11} & \bar{Q}_{12} & \bar{Q}_{13} \\ \bar{Q}_{21} & \bar{Q}_{22} & \bar{Q}_{23} \\ \bar{Q}_{31} & \bar{Q}_{32} & \bar{Q}_{33} \end{bmatrix} \begin{bmatrix} \varepsilon_x \\ \varepsilon_y \\ \gamma_{xy} \end{bmatrix} \qquad (4.95)$$

is obtained where

$$\bar{Q}_{11} = Q_{11}\cos^4\theta + Q_{22}\sin^4\theta + 2(Q_{12} + 2Q_{33})\sin^2\theta\cos^2\theta$$

$$\bar{Q}_{12} = \bar{Q}_{21} = (Q_{11} + Q_{22} - 4Q_{33})\sin^2\theta\cos^2\theta \\ + Q_{12}(\sin^4\theta + \cos^4\theta)$$

$$\bar{Q}_{13} = \bar{Q}_{31} = (Q_{11} - Q_{12} - 2Q_{33})\sin\theta\cos^3\theta \\ + (Q_{12} - Q_{22} + 2Q_{33})\sin^3\theta\cos\theta$$

$$\bar{Q}_{22} = Q_{11}\sin^4\theta + Q_{22}\cos^4\theta + 2(Q_{12} + 2Q_{33})\sin^2\theta\cos^2\theta$$

$$\bar{Q}_{23} = \bar{Q}_{32} = (Q_{11} - Q_{12} - 2Q_{33})\sin^3\theta\cos\theta \\ + (Q_{12} - Q_{22} + 2Q_{33})\sin\theta\cos^3\theta$$

$$\bar{Q}_{33} = (Q_{11} + Q_{22} - 2Q_{12} - 2Q_{33})\sin^2\theta\cos^2\theta \\ + Q_{33}(\sin^4\theta + \cos^4\theta)$$

and from eqns. (4.32),

$$Q_{11} = \frac{E_1}{1 - v_{12}v_{21}}$$

$$Q_{22} = \frac{E_2}{1 - v_{12}v_{21}}$$

$$Q_{12} = \frac{v_{21}E_1}{1 - v_{12}v_{21}} = \frac{v_{12}E_2}{1 - v_{12}v_{21}}$$

$$Q_{33} = G_{12}$$

The most important aspect of these compliance and material matrices is that there exist no zero terms, implying that the normal and shearing effects are no longer independent of each other.

Hence, referring to Fig. 4.11a, a direct stress in the x-direction would induce not only normal strains in the x- and y-directions but also a shearing strain with respect to the x–y axes. Such deformation, typical of completely anisotropic materials, is depicted in Fig. 4.16. The difference

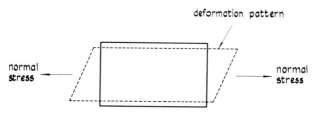

FIG. 4.16. Deformation of an anisotropic lamina under a uniaxial normal stress.

between a completely anisotropic lamina and an orthotropic lamina in which the principal material axes and the stress–strain coordinates are not aligned lies in the fact that in the former case six material constants would be required for a complete description of the stress–strain behaviour, whereas in the latter only four are required to quantify every element of the compliance and material matrices.

4.4 STRENGTH CHARACTERISTICS OF COMPOSITE LAMINAE

In addition to an understanding of the stiffness characteristics, defined by the stress–strain relationships developed in the previous section, a knowledge of the strengths, i.e. the magnitudes of the stresses that cause failure, of laminae is also essential for a comprehensive description of the characteristics of composite materials. Whereas the stress–strain relationships may be considered to describe the actual stresses occurring at any point in a lamina, the strength characteristics may be thought of as describing the allowable stresses at the point.

In considering the stiffness characteristics of a lamina, properties equal in both tension and compression were assumed. However, when investigating the strengths of materials, it is often found that in tension these characteristics are quite different from those in compression. Not only may a material have these varying failure characteristics, but also failure may occur by the rapid onset of some non-linear stress–strain behaviour, typified by the yielding of ductile materials such as mild steel, or it may occur suddenly during linear, or very nearly linear, behaviour, for example in brittle materials such as cast iron.

The ultimate behaviour of fibre composite materials may vary between these two extremes, depending upon the nature and properties of the constituent materials, as well as exhibiting different tensile and compressive properties; hence a systematic development of the strengths of these materials, such as has been carried out for stiffness properties, is not possible.

What can be achieved, however, is a development of a number of hypotheses regarding the prediction of strengths,[8] and the applicability of each to particular forms of composite can be investigated.

4.4.1 The Basic Concepts in the Determination of Strength Characteristics

The various relationships between stress and strain developed in section 4.3 were derived from the fundamental elastic material properties of Young's modulus, E, Poisson's ratio, v, and the modulus of rigidity, G.

In the isotropic case the material was defined by its single values of Young's modulus and Poisson's ratio, whereas in the two-dimensional orthotropic case, two values of Young's modulus, together with a Poisson's ratio value and a modulus of rigidity were required to quantify the material stiffness properties. In the three-dimensional case, of course, a correspondingly greater number of properties were required.

With a knowledge of these fundamental elastic material properties, which could be furnished from relatively simple experiments,[4,9–12] the complex deformation patterns of bodies subjected to two- or three-dimensional combined stress situations could be described.

It is this same philosophy which is the basis for the investigation of the strengths of materials. Just as the stiffness characteristics were found in terms of the elastic material properties, in the orthotropic case these properties being defined in the principal directions of the material, so it may be possible to obtain a guide to the strength of materials subjected to combined stresses in terms of a number of basic strengths which could be obtained from simple uniaxial tension and compression tests.[13] Again, in the orthotropic case, these basic strengths could be related to the principal material axes.

The development of these strength hypotheses will be considered in three dimensions, the results for two-dimensional behaviour being then obtained by neglecting stresses in one of the three directions.

4.4.2 Strength Hypotheses for Isotropic Laminae

Uniaxial tests on specimens of an isotropic material in tension or compression will furnish not only the Young's modulus and Poisson's ratio values but also information regarding the strength of the material, i.e. a knowledge of the stresses that will cause failure. However, it is not necessary that the applied stress causing failure in, for example, a tensile test produces a failure in tension, for during such a test shearing stresses are produced which may cause the failure to be one of shear. This may be shown by a consideration of the uniaxial tensile test depicted in Fig. 4.17a,b.

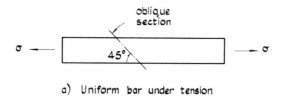

a) Uniform bar under tension

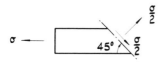

b) Normal and shearing stresses on oblique section

FIG. 4.17. Uniaxial tensile test.

If an oblique section through the bar is considered, then it is seen that for equilibrium of the portion depicted in Fig. 4.17b, both a normal and a shearing stress must act on the section. The normal stress is always less than the applied axial stress, σ, since σ is the maximum principal stress, and the shearing stress reaches a maximum value of $\sigma/2$ on a section inclined at 45° to the longitudinal axis of the bar. Thus the application of a tensile stress σ produces shearing stresses in the bar, the maximum value of these being $\sigma/2$. Similarly an axial compressive stress σ will produce a shearing stress of the same magnitude but acting in the opposite sense. Hence failure of the bar, whether this failure be an actual fracture or a yield, that is a distinct departure from the linear stress–strain curve may be due to either normal or shearing stress. Generally it is found that in ductile materials, i.e. those failing by yielding, the tensile strength is higher than the shearing strength and that in a tensile test failure due to shear often occurs, whereas in brittle materials, i.e. those failing by fracture, the tensile strength is very low in comparison with the shearing strength and hence in a tensile test failure in tension occurs. In a uniaxial compression test failure usually occurs due to shear (assuming that the specimen is of such proportions that a buckling failure cannot occur) since the shearing strength tends to be much lower than the compressive strength. Thus in a ductile material the applied failure stresses in both tensile and compressive tests will often be similar as both can induce shear failure, whereas in a brittle material the applied compressive failure stress will be higher than the corresponding tensile stress, since the former will cause failure in shear and the latter failure in tension.

Although both normal or shear failure in materials can occur, the basic experimentally determined strengths in which all the failure hypotheses are expressed are the applied tensile and compressive stresses that cause failure in uniaxial tests, irrespective of whether that failure is tensile or shear.

4.4.2.1 MAXIMUM PRINCIPAL STRESS THEORY

Consider a small three-dimensional element subjected to the three principal stresses σ_x, σ_y and σ_z, as shown in Fig. 4.6a; that is, there are no shearing stresses acting on the faces of the element.

If $\sigma_x > \sigma_y > \sigma_z$ then this hypothesis, usually attributed to Rankine, states that if σ_x is tensile, failure will occur when σ_x, the maximum principal stress, becomes equal to the failure stress, σ_{ut}, in a uniaxial tensile test, i.e.

$$\sigma_x = \sigma_{ut} \qquad (4.96)$$

Similarly if the minimum principal stress, σ_z, is compressive, failure can occur when

$$\sigma_z = \sigma_{uc} \qquad (4.97)$$

where σ_{uc} is the failure stress in a uniaxial compressive test.

4.4.2.2 MAXIMUM PRINCIPAL STRAIN THEORY

This hypothesis, first proposed by St-Venant, is similar to the maximum principal stress theory, but considers the possibility of failure occurring when the strain reaches ultimate proportions.

Assuming that failure occurs when the maximum principal tensile strain, ε_x, equals the tensile strain at failure in a uniaxial tensile test, ε_{ut}, then

$$\varepsilon_x = \varepsilon_{ut} \qquad (4.98)$$

which on expressing in terms of stresses yields

$$\frac{1}{E}(\sigma_x - \nu\sigma_y - \nu\sigma_z) = \frac{\sigma_{ut}}{E} \qquad (4.99)$$

that is

$$\sigma_x - \nu(\sigma_y + \sigma_z) = \sigma_{ut} \qquad (4.100)$$

Similarly, if the minimum principal strain, ε_z, is compressive and equal to ε_{uc}, the compressive strain at failure in a uniaxial compressive test, it ensues that

$$\varepsilon_z = \varepsilon_{uc} \qquad (4.101)$$

and hence

$$\sigma_z - \nu(\sigma_x + \sigma_y) = \sigma_{uc} \qquad (4.102)$$

When Poisson's ratio is zero, eqns. (4.100) and (4.102) become identical to eqns. (4.96) and (4.97) given by the maximum principal stress theory.

Since these two hypotheses consider failure due to normal stresses, irrespective of the magnitude of any shearing stresses present, they should be expected to find their application in the prediction of the failure of brittle materials when tension is present. Experimental investigations have indeed verified this deduction, the maximum principal stress theory having generally been found the most appropriate.

Ductile materials or brittle materials in compression, on the other hand, often display shear failure characteristics, so for such materials it would appear logical to consider criteria based on shearing stresses or strains in preference to those based on normal ones.

4.4.2.3 MAXIMUM SHEARING STRESS THEORY

Considering again Fig. 4.6a, it may be noted from a Mohr's stress circle[7,13,14] construction that in the x–z plane, i.e. the plane containing the maximum and minimum principal stresses, σ_x and σ_z respectively, the magnitude of the maximum shearing stress is

$$\tau_{max} = \frac{\sigma_x - \sigma_z}{2} \qquad (4.103)$$

and that this is the greatest shearing stress that can occur in the element.

Thus, assuming that the body fails in shear, it follows that failure occurs when the maximum shearing stress equals the shearing stress occurring in a uniaxial tensile test at failure, that is

$$\tau_{max} = \frac{\sigma_{ut}}{2} \qquad (4.104)$$

Thus

$$\frac{\sigma_x - \sigma_z}{2} = \frac{\sigma_{ut}}{2} \qquad (4.105)$$

or

$$\sigma_x - \sigma_z = \sigma_{ut} \qquad (4.106)$$

Hence failure occurs when the difference between the maximum and minimum principal stresses equals the failure stress in a uniaxial tensile test. It may be noticed that a maximum shearing strain theory will produce exactly the same result.

Being based on a shear failure this hypothesis, associated with the name of Tresca, may be expected to apply to those materials exhibiting weakness in shear, such as the ductile materials, and indeed for such materials it does give quite favourable results.

The hypotheses considered so far have expressed the criteria for failure as functions of either limiting stresses or limiting strains. Although both assumptions appear logically defensible, it is seen that in the case of the principal stress and strain theories, differing results are obtained. The question thus arises whether it is possible to combine the stress and strain criteria, so that criteria based on energy principles are generated. This idea is developed in the following discussion.

4.4.2.4 TOTAL STRAIN ENERGY THEORY

First proposed by Beltrami, but also developed by Haigh, this hypothesis assumes that at failure the strain energy per unit volume in the stressed body is equal to that per unit volume in a specimen under uniaxial tension at incipient failure.

Assuming linear elasticity, in a uniaxial tensile test the work done by the external force P in extending the specimen an amount u is

$$\text{Work done} = \frac{1}{2} P u \tag{4.107}$$

This is equal to the strain energy stored, and thus if the specimen is of length L and uniform cross-sectional area A, the strain energy stored per unit volume is

$$U = \frac{1}{2} \frac{P}{A} \frac{u}{L} = \frac{1}{2} \sigma \varepsilon \tag{4.108}$$

σ and ε being the direct stress and strain in the specimen.
Since $\varepsilon = \sigma/E$, this may be written

$$U = \frac{1}{2E} \sigma^2 \tag{4.109}$$

and hence at incipient failure the strain energy stored per unit volume is

$$U = \frac{1}{2E} \sigma_{ut}^2 \tag{4.110}$$

Using the same procedure, the strain energy stored per unit volume in the body shown in Fig. 4.6a is

$$U = \frac{1}{2} \sigma_x \varepsilon_x + \frac{1}{2} \sigma_y \varepsilon_y + \frac{1}{2} \sigma_z \varepsilon_z \tag{4.111}$$

and since

$$\varepsilon_x = \frac{1}{E}(\sigma_x - \nu \sigma_y - \nu \sigma_z)$$

$$\varepsilon_y = \frac{1}{E}(\sigma_y - \nu \sigma_z - \nu \sigma_x)$$

$$\varepsilon_z = \frac{1}{E}(\sigma_z - \nu \sigma_x - \nu \sigma_y)$$

this may be expressed as

$$U = \frac{1}{2E}(\sigma_x^2 + \sigma_y^2 + \sigma_z^2) - \frac{\nu}{2E}(2\sigma_x \sigma_y + 2\sigma_y \sigma_z + 2\sigma_z \sigma_x) \tag{4.112}$$

Thus on equating the two energy densities given in eqns. (4.110) and (4.112)

$$(\sigma_x^2 + \sigma_y^2 + \sigma_z^2) - 2\nu(\sigma_x\sigma_y + \sigma_y\sigma_z + \sigma_z\sigma_x) = \sigma_{ut}^2 \quad (4.113)$$

the stresses σ_x, σ_y, σ_z being the principal stresses in the x, y, z-directions respectively.

In a two-dimensional application, i.e. when for instance $\sigma_z = 0$, this result reduces to

$$\sigma_x^2 + \sigma_y^2 - 2\nu\sigma_x\sigma_y = \sigma_{ut}^2 \quad (4.114)$$

Now a very interesting observation may be made. If the body is subjected to compression then it may be thought that σ_{uc}^2 may be substituted in the right-hand sides of eqns. (4.113) or (4.114). This would imply, however, that the material failed in shear, since shearing strength is almost invariably lower than compressive strength and hence in a uniaxial compressive test the value of σ_{uc} represents a shear failure. However, if the body is subjected to equal principal compressive stresses, such as in a hydrostatic environment, it can easily be shown that no shearing stresses exist in the body, and the material experiences pure compression. In this situation, then, it would be expected that extremely large compressive stresses would be required to cause failure, and this has indeed been verified experimentally.

It is now seen that if in eqn. (4.113) all the principal stresses were made compressive and equal to σ, the equation would become

$$3(1 - 2\nu)\sigma^2 = \sigma_{uc}^2 \quad (4.115)$$

which would imply that the allowable applied compressive stress could be lower than σ_{uc}, a situation which is illogical.

The application of equal tensile stresses to the body similarly produces no shearing stress, and thus the theory appears more applicable to brittle materials, i.e. to materials in which the tensile strength is very low.

In considering a failure theory based on the energy concept applicable to ductile materials, i.e. materials whose lowest strength is that of shear, these 'hydrostatic' situations must be accounted for. It is this consideration that has led to the following successful hypothesis.

4.4.2.5 DISTORTION STRAIN ENERGY THEORY
It has been shown that enormous hydrostatic compressive stresses can be sustained by a body without failure occurring, and in ductile materials similar tensile stress situations may also be withstood.

Now equality of stresses, both in magnitude and sense, in three directions produces equality of strains and hence, while there is a change in volume of the body, there is no change of shape. Thus it would appear that failure of a ductile material is associated only with the change of shape or distortion of the body under stress and not with the change in volume.

Huber, in 1904, assumed that only the strain energy of distortion per unit volume be used to determine the failure criterion, this value being equated to the distortional strain energy per unit volume in a uniaxial tensile test at incipient failure. The concept was also independently considered by Maxwell, Hencky and von Mises, and is now commonly known as the von Mises criterion.

Consider again the body shown in Fig. 4.6a. If each of the principal stresses σ_x, σ_y and σ_z is expressed as the sum of two components thus:

$$\sigma_x = \sigma_v + \bar{\sigma}_x$$
$$\sigma_y = \sigma_v + \bar{\sigma}_y \quad (4.116)$$
$$\sigma_z = \sigma_v + \bar{\sigma}_z$$

then it may be deduced that when the body is subjected to the σ_v components, only a change of volume takes place, while under the components $\bar{\sigma}_x$, $\bar{\sigma}_y$, $\bar{\sigma}_z$ only a distortion of the body occurs. These two stress systems are shown in Fig. 4.18.

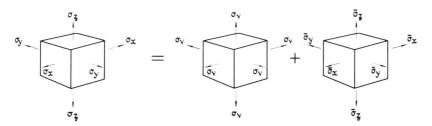

FIG. 4.18. 'Volume' and 'distortion' stress systems.

The values of these stress components can be obtained by noting that if the distortion components produce no change in volume the sum of the strains due to these components must be zero for small displacements.
Thus

$$\bar{\varepsilon}_x + \bar{\varepsilon}_y + \bar{\varepsilon}_z = 0 \quad (4.117)$$

which upon writing in terms of stresses becomes

$$\frac{1-2v}{E}(\bar{\sigma}_x+\bar{\sigma}_y+\bar{\sigma}_z)=0 \qquad (4.118)$$

that is,

$$\bar{\sigma}_x+\bar{\sigma}_y+\bar{\sigma}_z=0$$

which upon substituting into the relationships given in eqns. 4.116 produces

$$\sigma_x+\sigma_y+\sigma_z=3\sigma_v \qquad (4.119)$$

or

$$\sigma_v=\frac{\sigma_x+\sigma_y+\sigma_z}{3} \qquad (4.120)$$

This stress will thus produce strains, ε_v, in all three directions of magnitude

$$\varepsilon_v=\frac{1}{E}(\sigma_v-v\sigma_v-v\sigma_v)$$

$$=\frac{1-2v}{E}\sigma_v \qquad (4.121)$$

Now the total strain energy per unit volume in the body, U, is composed of the strain energy per unit volume due to the change in volume, U_v, together with that due to distortion, U_D.
Thus

$$U_D=U-U_v \qquad (4.122)$$

From eqn. (4.112), the total strain energy per unit volume is

$$U=\frac{1}{2E}(\sigma_x^2+\sigma_y^2+\sigma_z^2)-\frac{v}{2E}(2\sigma_x\sigma_y+2\sigma_y\sigma_z+2\sigma_z\sigma_x)$$

while from eqns. (4.111) and (4.121),

$$U_v=\frac{1}{2}\sigma_v\varepsilon_v+\frac{1}{2}\sigma_v\varepsilon_v+\frac{1}{2}\sigma_v\varepsilon_v$$

$$=\frac{3(1-2v)}{2E}\sigma_v^2 \qquad (4.123)$$

Hence from eqn. (4.122), and using eqn. (4.120), the strain energy of distortion per unit volume is given by

$$U_D = \frac{1}{2E}(\sigma_x^2 + \sigma_y^2 + \sigma_z^2) - \frac{v}{2E}(2\sigma_x\sigma_y + 2\sigma_y\sigma_z + 2\sigma_z\sigma_x)$$

$$-\frac{(1-2v)}{6E}(\sigma_x + \sigma_y + \sigma_z)^2$$

$$= \frac{1+v}{6E}[(\sigma_x - \sigma_y)^2 + (\sigma_y - \sigma_z)^2 + (\sigma_z - \sigma_x)^2] \quad (4.124)$$

The value of the distortional strain energy per unit volume in a uniaxial tensile test at failure may be obtained from eqn. (4.124) by letting σ_y and σ_z be zero and replacing σ_x by σ_{ut}.
Thus

$$U_D = \frac{1+v}{6E}2\sigma_{ut}^2 \quad (4.125)$$

Hence equating the right-hand expressions in eqns. (4.124) and (4.125) produces the criterion

$$(\sigma_x - \sigma_y)^2 + (\sigma_y - \sigma_z)^2 + (\sigma_z - \sigma_x)^2 = 2\sigma_{ut}^2 \quad (4.126)$$

In a two-dimensional system one of the stresses, say σ_z, is zero and the equation then becomes

$$\sigma_x^2 + \sigma_y^2 - \sigma_x\sigma_y = \sigma_{ut}^2 \quad (4.127)$$

This criterion has proved to be the most satisfactory hypothesis in predicting the behaviour of ductile materials.

It may be noticed from eqns. (4.106) and (4.126) that this theory and the maximum shearing stress theory, which also predicts the failure of ductile materials with reasonable accuracy, both give criteria that are based on principal stress differences.

In attempting to apply a strength hypothesis to glass fibre-reinforced composites, it must be noted that such materials may exhibit varying degrees of strength and ductility depending upon the amount and orientation of the fibres, and hence the selection of a particular strength theory is not obvious. However, it is considered that for glass-reinforced plastics that may be classed as isotropic, the distortional strain energy theory is a criterion that can be applied with an acceptable degree of confidence.

4.4.3 Strength Hypotheses for Orthotropic Laminae

The strength hypotheses for orthotropic materials must, as in the case of isotropic materials, be based on the results of simple fundamental tests. Since an orthotropic material has different strengths in different directions, this means that a more extensive set of basic test data is required than for isotropic materials. These basic results are most fundamental to orthotropic materials when measured in the principal directions of the material. Thus three uniaxial tests are required, one in each of the principal material directions, to determine the three values of Young's modulus and the associated Poisson's ratio results, and also the strengths in these directions. Uniaxial tests of this nature, i.e. in the directions of the principal axes of the material, eliminate any coupling of shearing and normal strains that would occur if the stresses were applied in any other directions, and thus produce only normal strains relative to these directions.

In the case of isotropic materials, it has been shown that the shearing strength can be obtained from the uniaxial tests. However, in orthotropic materials, the basic shearing strengths with respect to the principal material directions must be determined from independent experiments in which only shearing strains are induced with respect to the principal material axes.

These basic tests are shown diagrammatically in two-dimensional form in Fig. 4.19.

Because the basic strengths are defined in the fixed directions of the principal material axes, irrespective of the actual state of stress in the material, the use of maximum principal stresses and maximum shearing stresses, so fundamental to the treatment of isotropic materials, becomes of little value in determining the allowable stresses in orthotropic materials.[3] Therefore, instead of considering the stresses at a point to be resolved to give their maximum values, irrespective of direction, the stress condition at a point is resolved into its normal and shearing components relative to the principal material axes at the point, and thus the failure criteria in orthotropic materials become functions of the basic normal and shearing strengths described previously.

With this observation of the difference between the strengths of isotropic and orthotropic materials, the corresponding strength hypotheses for orthotropic materials can now be developed.

4.4.3.1 MAXIMUM STRESS THEORY
Analogously to the maximum principal stress and maximum shearing

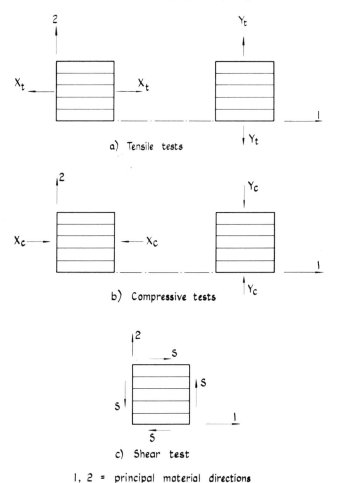

FIG. 4.19. Basic strength and stiffness tests for an orthotropic material in two dimensions.

stress theories for isotropic materials, this hypothesis suggests that if failure does not occur, then the stresses in the principal material directions are less than the corresponding strengths as determined by the basic tests.

Thus, considering the stresses in two dimensions, and observing again Fig. 4.19, the following conditions must apply for tensile stresses if failure is not to occur:

$$\sigma_1 < X_t$$
$$\sigma_2 < Y_t \quad (4.128a)$$
$$|\tau_{12}| < S$$

and for compressive stresses:

$$|\sigma_1| < |X_c|$$
$$|\sigma_2| < |Y_c| \quad (4.128b)$$
$$|\tau_{12}| < S$$

where X_t and X_c are the tensile and compressive strengths in the 1-direction, Y_t and Y_c are the tensile and compressive strengths in the 2-direction and S is the shearing strength in the 1–2 plane. Thus this hypothesis requires that not one but three criteria be satisfied.

4.4.3.2 MAXIMUM STRAIN THEORY

This hypothesis is related to the maximum stress theory in the same manner as the maximum principal stress and strain theories are related for isotropic materials. Thus for tensile strains in two dimensions,

$$\varepsilon_1 < \varepsilon_{x_t}$$
$$\varepsilon_2 < \varepsilon_{y_t} \quad (4.129a)$$
$$|\gamma_{12}| < \varepsilon_s$$

while for compressive strains,

$$|\varepsilon_1| < |\varepsilon_{x_c}|$$
$$|\varepsilon_2| < |\varepsilon_{y_c}| \quad (4.129b)$$
$$|\gamma_{12}| < \varepsilon_s$$

where ε_{x_t} and ε_{x_c} are the maximum tensile and compressive strains in the 1-direction, ε_{y_t} and ε_{y_c} are the maximum tensile and compressive strains in the 2-direction and ε_s is the maximum shearing strain in the 1–2 plane.

These criteria may, of course, be stated in terms of stresses by the application of the stress–strain equations. Thus eqns. (4.129a) would become, assuming no stress in the 3-direction,

$$\sigma_1 - v_{12}\sigma_2 < X_t$$
$$\sigma_2 - v_{21}\sigma_1 < Y_t \qquad (4.130)$$
$$|\tau_{12}| < S$$

and eqns. (4.129b) may be similarly transformed.

These two hypotheses give fairly reasonable predictions for the failure of orthotropic glass fibre-reinforced composites, although in the majority of cases a more satisfactory agreement with experimental results is provided by the following energy criterion.

4.4.3.3 TSAI–HILL ENERGY THEORY

When developing the distortional strain energy theory for isotropic materials, the normal stresses acting on the body were considered to be the principal stresses. If these normal stresses were not the principal ones, then the body would also be subjected to the three shearing stresses τ_{xy}, τ_{yz}, τ_{zx} on the faces of the body, and these would produce the additional strain energy

$$\frac{1}{2}\tau_{xy}\gamma_{xy} + \frac{1}{2}\tau_{yz}\gamma_{yz} + \frac{1}{2}\tau_{zx}\gamma_{zx}$$

which on substitution of the stress–strain relationships

$$G = \frac{E}{2(1+v)} = \frac{\tau_{xy}}{\gamma_{xy}} = \frac{\tau_{yz}}{\gamma_{yz}} = \frac{\tau_{zx}}{\gamma_{zx}}$$

becomes

$$\frac{1+v}{E}(\tau_{xy}^2 + \tau_{yz}^2 + \tau_{zx}^2)$$

Since shearing strains only distort the element and cause no change of volume, the total distortional strain energy density becomes, by extending eqn. (4.124),

$$U_D = \frac{1+v}{6E}[(\sigma_x - \sigma_y)^2 + (\sigma_y - \sigma_z)^2 + (\sigma_z - \sigma_x)^2]$$
$$+ \frac{1+v}{E}(\tau_{xy}^2 + \tau_{yz}^2 + \tau_{zx}^2) \qquad (4.131)$$

Equating this to the value of the distortional strain energy density at failure in a uniaxial tensile test given in eqn. (4.125), i.e. to

$$\frac{1+v}{6E}2\sigma_{ut}^2$$

and dividing both sides by this value gives

$$\frac{1}{2\sigma_{ut}^2}[(\sigma_x-\sigma_y)^2+(\sigma_y-\sigma_z)^2+(\sigma_z-\sigma_x)^2]$$
$$+\frac{3}{\sigma_{ut}^2}(\tau_{xy}^2+\tau_{yz}^2+\tau_{zx}^2)=1 \quad (4.132)$$

It is to be remembered that the normal stresses in this equation are not in general principal stresses.

If this form of failure criterion is assumed to apply to orthotropic materials, then the anologous equation must be expressed in terms of the stresses in the principal material directions together with the three normal strengths X, Y, Z and the three corresponding shearing strengths, thus producing a criterion of the form

$$\bar{F}(\sigma_1-\sigma_2)^2+\bar{G}(\sigma_2-\sigma_3)^2+\bar{H}(\sigma_3-\sigma_1)^2$$
$$+\bar{L}\tau_{12}^2+\bar{M}\tau_{23}^2+\bar{N}\tau_{31}^2=1 \quad (4.133)$$

where the six values $\bar{F}, \bar{G}, \bar{H}, \bar{L}, \bar{M}, \bar{N}$ are obtained from the six basic test results.[3,15,16]

Thus on rewriting eqns. (4.133) in the form

$$(\bar{F}+\bar{H})\sigma_1^2+(\bar{F}+\bar{G})\sigma_2^2+(\bar{G}+\bar{H})\sigma_3^2$$
$$-2\bar{F}\sigma_1\sigma_2-2\bar{G}\sigma_2\sigma_3-2\bar{H}\sigma_3\sigma_1 \quad (4.134)$$
$$+\bar{L}\tau_{12}^2+\bar{M}\tau_{23}^2+\bar{N}\tau_{31}^2=1$$

it is seen that the application of σ_1 only results in

$$\bar{F}+\bar{H}=\frac{1}{X^2}$$

Similarly

$$\bar{F}+\bar{G}=\frac{1}{Y^2}$$

and

$$\bar{G}+\bar{H}=\frac{1}{Z^2}$$

where Z is the strength in the direction normal to the 1–2 plane, i.e. in the 3-direction.
Hence

$$2\bar{F} = \frac{1}{X^2} + \frac{1}{Y^2} - \frac{1}{Z^2}$$

$$2\bar{G} = \frac{1}{Y^2} + \frac{1}{Z^2} - \frac{1}{X^2} \qquad (4.135)$$

$$2\bar{H} = \frac{1}{Z^2} + \frac{1}{X^2} - \frac{1}{Y^2}$$

Similarly

$$\bar{L} = \frac{1}{S^2}$$

and $\bar{M}$ and $\bar{N}$ are given in terms of the shearing strengths relative to the 2–3 and 3–1 planes respectively.

For a lamina under plane stress in the 1–2 plane

$$\sigma_3 = \tau_{23} = \tau_{31} = 0$$

and further, if the lamina is unidirectionally reinforced in the 1-direction, it can be inferred from symmetry of the cross-section that the strengths Y and Z are equal.[3] Thus in this case the failure criterion becomes

$$\frac{\sigma_1^2}{X^2} + \frac{\sigma_2^2}{Y^2} - \frac{\sigma_1 \sigma_2}{X^2} + \frac{\tau_{12}^2}{S} = 1 \qquad (4.136)$$

The Tsai–Hill failure criterion is not based completely on distortion in the orthotropic case since a hydrostatic stress condition can, in such materials, produce distortion as well as volume change. However the theory assumes that such a volume change has no effect on failure, and the application of the criterion has shown good agreement with experimental results.[17]

This chapter has developed the concepts of stiffness and strength of laminae from the fundamental parameters of elastic constants and basic strengths. These fundamental properties have not yet been related to the properties of the constituents of the composite. It is this subject that must next be examined.

REFERENCES

1. TIMOSHENKO, S. P. and GOODIER, J. N., *Theory of Elasticity*, 3rd edn., 1970, McGraw-Hill, New York.
2. LIVESLEY, R. K., *Matrix Methods of Structural Analysis*, 1969, Pergamon Press, Oxford.
3. JONES, R. M., *Mechanics of Composite Materials*, 1975, McGraw-Hill, New York.
4. AGARWAL, B. D. and BROUTMAN, L. J., *Analysis and Performance of Fiber Composites*, 1980, John Wiley, New York.
5. ODEN, J., *Mechanics of Elastic Structures*, 1967, McGraw-Hill, New York.
6. SECHLER, E. E., *Elasticity in Engineering*, 1960, John Wiley, New York.
7. TIMOSHENKO, S. P., *Strength of Materials*, Part 1, 3rd edn., 1980, Van Nostrand Reinhold, New York.
8. KAMINSKI, B. E. and LANTZ, R. B., Strength theories of failure for anisotropic materials, in *Composite Materials, Testing and Design, Symposium, ASTM*, Feb. 1969.
9. FISCHER, S., ROMAN, I., HAREL, H., MAROM, G. and WAGNER, H. D., Simultaneous determination of Young's and shear moduli in composites, *ASTM J. Testing and Evaluation*, **9**, Sept. 1981, 303.
10. HOLMES, M. and AL-KHAYATT, Q. C., Structural properties of glass reinforced plastics, *Composites*, **6**, July 1975, 157.
11. VOLOSHIN, A. and ARCAN, M., Pure shear moduli of unidirectional fibre-reinforced materials (FRM), *Fibre Science and Technology*, **13**, 1980, 125.
12. YEOW, Y. T. and BRINSON, H. F., A comparison of simple shear characterisation methods for composite laminates, *Composites*, **9**, Jan. 1978, 49.
13. BENHAM, P. P. and WARNOCK, F. V., *Mechanics of Solids and Structures*, 1980, Pitman, London.
14. RYDER, G. H., *Strength of Materials*, 3rd edn., 1974, MacMillan, London.
15. HILL, R., *The Mathematical Theory of Plasticity*, 1950, Oxford University Press, London.
16. TSAI, S. W., Strength theories of filamentary structures, in *Fundamental Aspects of Fiber Reinforced Plastic Composites*, ed. R. T. Schwartz and H. S. Schwartz, 1968, Wiley Interscience, New York.
17. OWEN, M. J. and RICE, D. J., Biaxial strength behaviour of glass fabric reinforced polyester resins, *Composites*, **12**, Jan. 1981, 13.

CHAPTER 5

The Theoretical Micromechanical Analysis of Composite Laminae

5.1 INTRODUCTION

The stress–strain relationships for a single lamina developed in the previous chapter provide a vital component in the understanding of the mechanical behaviour of composite materials. However, it is not possible to use these relationships until the salient elastic properties of Young's moduli, Poisson's ratios and moduli of rigidity for the lamina are determined.

Of course it would be possible to obtain these properties experimentally but this would only produce the properties of the individual material so tested. From a design standpoint, the process of obtaining the most economical form of composite for a particular application requires methods of determining the ratios of glass fibre and resin, and also the form of fibre most suitable, to produce the required elastic properties.

Thus the complete design process would entail determining firstly the properties required by the composite material to fulfil the structural functions of adequate stiffness and strength, and secondly the quantity and configuration of the fibre reinforcement to produce such properties. The former macromechanical process has already been considered. The latter process to which attention must now be devoted, and which is akin to the determination of the amount and form of steel reinforcement required in reinforced concrete design, forms the study of the *micromechanical* analysis of the lamina.

5.2. ASSUMPTIONS AND LIMITATIONS IN MICROMECHANICAL ANALYSES

Before developing methods by which the structural properties of a composite may be assessed, it must first be recognised that the actual nature of a glass fibre-reinforced composite material is extremely complex, and that in order to produce tractable micromechanical solutions certain simplifying assumptions as to its properties are necessary.

The reasons for the various assumptions that must be made can readily be appreciated by considering the ideals aimed for in the manufacture of composites and recognising the limitations inherent in realising them.

The two basic properties which are aimed for in composite design are the maximum stiffness and strength for a given form of material, which is almost always assumed to possess linear elasticity. In order to approach these ideal properties, it is required that the material should be as uniform as possible so as to decrease the magnitude and number of weaker areas which could reduce the structural efficiency of the material.

This uniformity in material properties may be more nearly realised if firstly both the fibres and the matrix have uniform stiffness and strength characteristics. Therefore the resin matrix should be uniformly and completely cured, and the fibres should be, as far as possible, free from flaws which could produce strength variations.

Secondly, a high degree of uniformity can be achieved by careful manufacture of the composite. In this context the number of voids in the resin should be reduced as far as possible so that a high degree of bonding is produced between the matrix and the fibres. Also uniform spacing and perfect alignment of the fibres should be aimed for in the case of continous fibre reinforcement, and in the case of random chopped strands, no directional bias should be present.

These ideals lead to the following assumptions that are often made in the micromechanical analysis of a composite:

(1) The matrix and fibres are both homogeneous and isotropic.
(2) The matrix and fibres, and hence the composite, exhibit linear elastic behaviour.
(3) The fibres are uniform, regularly spaced, and perfectly aligned.
(4) The composite is macroscopically homogenous, with no voids, and perfect bonding exists between the fibres and the resin matrix.

Obviously, in practice, such ideal properties can never be attained, so the validity of results obtained from solutions based on these assumptions must be tested by experiment.

A number of approaches have been developed in the study of micromechanical behaviour, these being based either on the theory of elasticity or on the simpler idealisation afforded by a strength of materials approach.[1-3] Simpler results, possibly more amenable to immediate design use, are obtained from this latter procedure; it is this form of analysis that will be considered in this chapter.

5.3 THE STIFFNESS CHARACTERISTICS OF GLASS-REINFORCED LAMINAE

Glass fibre-reinforced laminae may be constructed as either continuously or discretely reinforced composites, continuous reinforcement forming unidirectional or bidirectional laminae possessing orthotropic properties, and random discrete reinforcement producing laminae which are essentially isotropic in character.

Although these three forms have obvious differences, the fundamental form of reinforcement used for the estimation of lamina stiffness properties is that composed of continuous unidirectional fibres. Estimation of the characteristics of the remaining two forms may be obtained by modification of the results thus obtained.

5.3.1 Stress–Strain Relationships in a Continuous Unidirectional Fibre Lamina

Consider the lamina shown in Fig. 5.1 which is reinforced in the 1-direction with continuous unidirectional fibres.

Considering the small rectangular element defined by the broken lines, it is observed that the lamina may be viewed as consisting of a number of such elements, each behaving in a macroscopically homogeneous manner. The properties of the lamina may be obtained by a consideration of the properties of such a representative element.[3]

Since the properties of the lamina will depend on the proportions of fibres and matrix present, any elastic property of the lamina may be expressed generally as:

$$C = C(E_f, v_f, V_f, E_m, v_m, V_m)$$

where the subscripts f and m refer to the fibres and matrix, respectively, E is Young's modulus, v is Poisson's ratio and V is proportion by volume

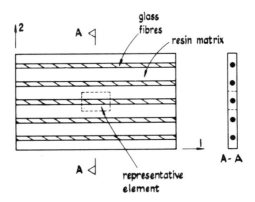

FIG. 5.1. Continuously and unidirectionally reinforced lamina.

in the composite. Hence expressions for the four elastic constants E_1, E_2, v_{12}, G_{12}, defining the properties of an orthotropic lamina, may be obtained in terms of these constituent properties, as will now be shown.

5.3.1.1 DETERMINATION OF YOUNG'S MODULUS, E_1

The determination of the value of Young's modulus, E_1, of the composite in the fibre direction, that is in the 1-direction, is based on the assumption that when the representative element of the lamina (Fig. 5.2) is subjected to stress in the 1-direction, the strains in the matrix and the fibre in this direction are identical.

With this assumption, the axial stress in the fibre, σ_f, and that in the matrix, σ_m, may be written as

$$\sigma_f = E_f \varepsilon_1 \tag{5.1a}$$

$$\sigma_m = E_m \varepsilon_1 \tag{5.1b}$$

where ε_1 is the longitudinal strain in the 1-direction, which is induced in both the fibre and the matrix.

Now, from equilibrium considerations, the resultant axial force, P, in the element is equal to the sum of the axial forces in the fibre and the matrix.
Hence

$$P = \sigma_f A_f + \sigma_m A_m \tag{5.2}$$

where A_f and A_m are the cross-sectional areas of the fibre and matrix respectively.

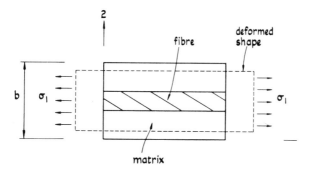

FIG. 5.2. Representative element under mean normal stress σ_1.

Now the resultant force, P, may be expressed in terms of a mean normal stress, σ_1, acting on the element thus:

$$P = \sigma_1 A \qquad (5.3)$$

where A is the cross-sectional area of the element.

Then eqn (5.2) may be rewritten as

$$\sigma_1 A = \sigma_f A_f + \sigma_m A_m \qquad (5.4)$$

which on dividing throughout by ε_1 and A gives

$$\frac{\sigma_1}{\varepsilon_1} = E_f \frac{A_f}{A} + E_m \frac{A_m}{A} \qquad (5.5)$$

Now σ_1/ε_1 is the macroscopic value of Young's modulus for the composite in the 1-direction, E_1, while A_f/A and A_m/A are V_f and V_m respectively.

Therefore equation 5.5 may be written as

$$E_1 = E_f V_f + E_m V_m \qquad (5.6)$$

where V_m may be expressed as $(1 - V_f)$.

Equation 5.6 may thus be written in the form

$$E_1 = (E_f - E_m) V_f + E_m \qquad (5.7)$$

and since E_m is much smaller than E_f, this shows clearly how dependent E_1 is upon the volume of fibres present. This relationship is shown pictorially in Fig. 5.3.

In using the assumption that the strains throughout both constituents

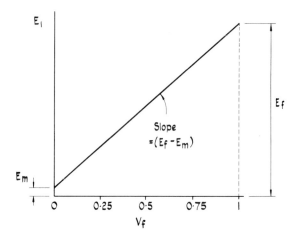

FIG. 5.3. Variation of E_1 with fibre content.

are equal, no shearing stresses will be induced along the fibre–matrix boundaries and the axial stresses in the fibres and matrix will be uniform, as stated in eqn (5.2).

Thus the derivations from the initial assumption of equal strains are consistent, and the simple relationship of eqn. 5.6 is, as expected, found to yield a close agreement with results obtained in practice.

5.3.1.2 Determination of Poisson's Ratio, v_{12}

An expression for v_{12} can be similarly deduced from a consideration of the element shown in Fig. 5.2 under the mean normal stress σ_1, again invoking the assumption of equal longitudinal strains in the fibres and matrix.

By definition, v_{12} is given as

$$v_{12} = -\frac{\varepsilon_2}{\varepsilon_1} \tag{5.8}$$

ε_2 being the lateral strain on the element in the 2-direction.

Now ε_2 is a function of two components, the lateral strain in the fibre and the lateral strain in the matrix, and these may be written respectively as

$$-v_f \varepsilon_1$$

and

$$-v_m \varepsilon_1$$

The respective lateral displacements in the 2-direction due to these strains are dependent on the geometry of the fibre within the matrix and also on the cross-sectional form of the representative element.

Assume first that the fibre is rectangular and extends over the whole thickness of the element of breadth b as depicted in Fig. 5.4a, and further neglect the effect of any strains that may occur in the constituents in the 3-direction.

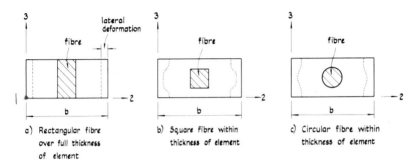

a) Rectangular fibre over full thickness of element

b) Square fibre within thickness of element

c) Circular fibre within thickness of element

NOTE :- Dotted lines represent extent of lateral deformation due to tensile strain in 1-direction

FIG. 5.4. Lateral strain in representative element due to uniform normal strain in 1-direction.

In this case the lateral displacement in the fibre may be written as

$$\text{Lateral strain in fibre} \times \text{width of fibre} = -v_f \varepsilon_1 \times V_f b \quad (5.9a)$$

while the lateral displacement in the matrix may be similarly expressed as

$$\text{Lateral strain in matrix} \times \text{width of matrix} = -v_m \varepsilon_1 \times V_m b \quad (5.9b)$$

Hence

$$\begin{aligned} \text{Total lateral displacement in the 2-direction} \\ = -(v_f \varepsilon_1 \times V_f b) - (v_m \varepsilon_1 \times V_m b) \\ = -(v_f V_f + v_m V_m) \varepsilon_1 b \end{aligned} \quad (5.9c)$$

Thus the lateral strain is

$$\varepsilon_2 = -(v_f V_f + v_m V_m) \varepsilon_1 \quad (5.10)$$

and hence

$$v_{12} = \frac{-\varepsilon_2}{\varepsilon_1} = v_f V_f + v_m V_m \qquad (5.11)$$

Again, by expressing V_m as $(1 - V_f)$, this result may be written

$$v_{12} = (v_f - v_m) V_f + v_m \qquad (5.12)$$

If the more realistic fibre reinforcement geometries shown in Figs 5.4b,c are considered, it can be noted that in these cases the lateral displacement of the composite element may vary across the thickness due to the difference that may exist between the values of Poisson's ratio in the fibre and matrix. If, for instance, Poisson's ratio for the matrix were greater than that for the fibre, then in the regions of the matrix above and below the fibre there would tend to be a greater displacement than in the region containing the fibre, and this variation in strain would cause shearing and normal stresses to be induced in the region of the fibre–matrix interface.

Corrections for these effects, and also perhaps for the effect of any differential strains in the 3-direction, may be considered through the more advanced theory of elasticity, but such complex considerations may be of doubtful value in view of the fact that a large number of analytically unquantifiable variables, such as non-alignment of fibres and imperfect bonding, also exist in a practical situation.

Also, although the values of Poisson's ratio for the matrix and fibre will generally be different, they will usually be of a similar order, somewhere within the middle of the isotropic range, and thus the effects referred to above should be of minor significance.

Therefore it may be considered reasonable to utilise eqn. (5.11) to determine Poisson's ratio for the unidirectional composite irrespective of fibre form. It may be noted that in so doing, the effect of the fibre is averaged over the thickness of the lamina, i.e. it is considered to be of the form shown in Fig. 5.4a.

It is found in practice that eqn. (5.11) does indeed predict the value of v_{12} with reasonable accuracy, and the manner in which this constant varies with V_f is shown in Fig. 5.5, where it is seen that a similar variation is attained to that for E_1, due of course to the identical form of eqns. (5.6) and (5.11).

5.3.1.3 DETERMINATION OF YOUNG'S MODULUS, E_2

A prediction of the transverse modulus of elasticity, E_2, may be obtained

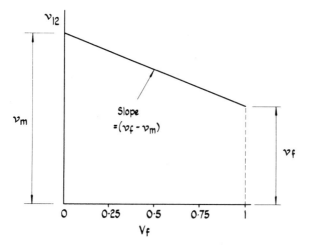

FIG. 5.5. Variation of v_{12} with fibre content.

from a consideration of the representative element subjected to a uniform normal stress, σ_2, on the boundaries in the 2-direction, as shown in Fig. 5.6a.

However, whereas the application of a uniform boundary strain in the 1-direction, used as the basis for the determination of E_1, subjects the element to the same longitudinal strain throughout, no similar constant stress condition occurs in the case of the element subjected to σ_2. That this is so is readily seen from an examination of the effect of Poisson's ratio.

Consider first the simplest case of a rectangular fibre extending over the thickness of the element, as shown in Fig. 5.4a. It may be noted that this fibre–matrix form was the one chosen for the derivation of the basic equation used to predict v_{12}.

If Poisson's ratio for both the fibre and matrix were zero, application of σ_2 would result in an absence of strain in the 1-direction. Hence only normal strain in the 2-direction would result, implying constancy of σ_2 throughout the matrix and fibre, the strains in these constituents being, of course, different. In reality Poisson's ratio is non-zero, and hence, since the fibre and matrix undergo differing strains in the 2-direction, due to the difference in their moduli of elasticity, they will exhibit differing strains in the 1-direction, as depicted in Fig. 5.6b. The ensuing displacement differential in the constituents induces varying shearing stresses in

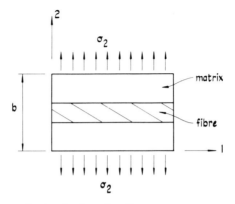

a) Application of uniform stress σ_2

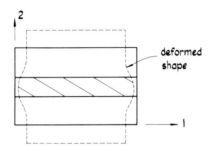

b) Deformation due to σ_2

FIG. 5.6. Representative element under uniform normal stress σ_2.

the 1–2 plane, these being particularly pronounced in the vicinity of the fibre–matrix interface, with the accompanying production of a variation in the normal stress distribution, thus violating any assumption as to the constancy of σ_2.

In progressing to the fibre shapes shown in Figs. 5.4b,c further complexities are introduced into the behaviour of the element, these having been described in a qualitative sense previously in discussing the prediction of v_{12}.

The simplest analysis for the determination of E_2 would therefore be one in which the Poisson effect is neglected, i.e. σ_2 is considered constant throughout the element, although it must be realised at the outset that such an analysis incorporates the inconsistencies described, and hence must be an approximation only. The following derivation based on the assumption of constancy of σ_2 is now considered for the prediction of E_2, the element having the section shown in Fig. 5.4a.

Due to the constant stress, σ_2, the transverse strains induced in the fibre and matrix are

$$\varepsilon_f = \frac{\sigma_2}{E_f} \qquad (5.13a)$$

and

$$\varepsilon_m = \frac{\sigma_2}{E_m} \qquad (5.13b)$$

respectively. As described in section 5.3.1.2, the transverse displacement in the fibre is given by:

Transverse strain in fibre × width of fibre $= \varepsilon_f \times V_f b$ (5.14a)

while the transverse displacement in the matrix is

Transverse strain in matrix × width of matrix $= \varepsilon_m \times V_m b$ (5.14b)

Thus

Total transverse displacement $= (\varepsilon_f \times V_f b) + (\varepsilon_m \times V_m b)$ (5.14c)

which by virtue of eqns. (5.13) equals

$$\sigma_2 b \left(\frac{V_f}{E_f} + \frac{V_m}{E_m} \right) \qquad (5.15)$$

and hence the transverse strain is

$$\varepsilon_2 = \sigma_2 \left(\frac{V_f}{E_f} + \frac{V_m}{E_m} \right) \qquad (5.16)$$

Now $E_2 = \sigma_2/\varepsilon_2$, and hence

$$E_2 = \frac{E_f E_m}{E_m V_f + E_f V_m} \qquad (5.17a)$$

$$= \frac{E_f E_m}{E_m V_f + E_f (1 - V_f)} \qquad (5.17b)$$

It may also be noted that eqns. (5.17) can be expressed in the form

$$\frac{1}{E_2} = \frac{V_f}{E_f} + \frac{V_m}{E_m} \tag{5.18a}$$

$$= \frac{V_f}{E_f} + \frac{(1-V_f)}{E_m} \tag{5.18b}$$

The manner in which E_2 depends on the volume of fibres in the composite is shown in Fig. 5.7, where it is noticed that for a volume of fibres as large as 50% the effect of the fibres only increases the transverse

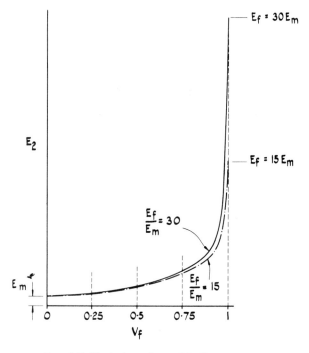

FIG. 5.7. Variation of E_2 with fibre content.

modulus of elasticity, E_2, to approximately twice the matrix modulus E_m, even when E_f is as many as 30 times the value of E_m. It is not until much higher volumes of fibre are present that the fibres themselves have any substantial effect in raising the value of the transverse modulus. This is in contrast to the effect of the fibres on the longitudinal modulus, E_1.

5.3.1.4 DETERMINATION OF MODULUS OF RIGIDITY, G_{12}

Just as the derivations of E_1 and v_{12} are formally the same, so the derivations of G_{12} is analogous to that for E_2, the representative element being considered to be subjected to uniform shearing and complementary shearing stresses along the boundaries, as shown in Fig. 5.8a.

Under this stress system, and due to the variation in shearing stiffness between the fibres and the matrix, an element with the idealised cross-section shown in Fig. 5.4a will deform as depicted in Fig. 5.8b.

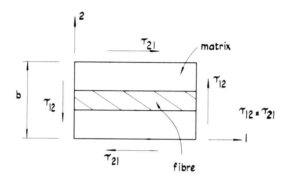

a) Application of uniform stresses τ_{12} and τ_{21}

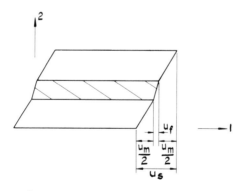

b) Deformation due to τ_{12} and τ_{21}

FIG. 5.8. Representative element under uniform shearing stresses τ_{12} and τ_{21}.

The shearing strains induced in the fibre and the matrix are given by

$$\gamma_f = \frac{\tau_{12}}{G_f} \tag{5.19a}$$

and

$$\gamma_m = \frac{\tau_{12}}{G_m} \tag{5.19b}$$

respectively, and the corresponding displacements are hence

Shearing displacement in fibre = shearing strain in fibre × width of fibre

$$= \gamma_f \times V_f b = u_f \tag{5.20a}$$

and

Shearing displacement in matrix = shearing strain in matrix × width of matrix

$$= \gamma_m \times V_m b = u_m \tag{5.20b}$$

The total shearing displacement, u_s, is thus given by

$$u_s = u_f + u_m = (\gamma_f \times V_f b) + (\gamma_m \times V_m b) \tag{5.21}$$

or, on substituting eqns. (5.19),

$$u_s = \tau_{12} b \left(\frac{V_f}{G_f} + \frac{V_m}{G_m} \right) \tag{5.22}$$

Now, if it is assumed that the shearing strain in the complete element is given by

$$\gamma_{12} = \frac{u_s}{b}$$

then from eqn. (5.22),

$$\gamma_{12} = \tau_{12} \left(\frac{V_f}{G_f} + \frac{V_m}{G_m} \right)$$

$$= \tau_{12} \left(\frac{G_m V_f + G_f V_m}{G_f G_m} \right) \tag{5.23}$$

Hence since

$$G_{12} = \frac{\tau_{12}}{\gamma_{12}}$$

it follows that

$$G_{12} = \frac{G_f G_m}{G_m V_f + G_f V_m} \quad (5.24a)$$

$$= \frac{G_f G_m}{G_m V_f + G_f(1-V_f)} \quad (5.24b)$$

which is of the same form as the expression for E_2.

Again eqns. (5.24) may be expressed in the form

$$\frac{1}{G_{12}} = \frac{V_f}{G_f} + \frac{V_m}{G_m} \quad (5.25a)$$

$$= \frac{V_f}{G_f} + \frac{(1-V_f)}{G_m} \quad (5.25b)$$

The way in which G_{12} varies with the volume of fibres is shown in Fig. 5.9 where it is seen that, like E_2, it is the modulus of rigidity of the matrix that determines the order of G_{12} unless the volume of fibres becomes exceptionally high.

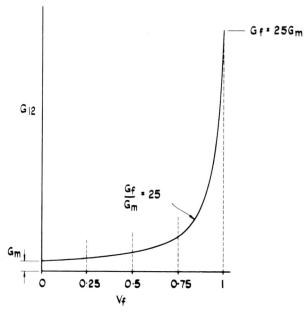

FIG. 5.9. Variation of G_{12} with fibre content.

5.3.1.5 Discussion

The results presented in subsections 5.3.1 to 5.3.4 are of necessity based on simplifying assumptions and idealisations, and therefore must be viewed as approximations to the actual behaviour of such composites.

In spite of these assumptions, however, the results for E_1, the value of Young's modulus in the fibre direction, and for the Poisson's ratio value, v_{12}, show very good agreement with experiment, and demonstrate that the properties are highly dependent on the percentage of fibres present. The remaining two properties, E_2, the transverse Young's modulus, and G_{12}, the modulus of rigidity with respect to the 1- and 2-directions, depend on the fibres to a much smaller degree, their values being more dependent on the matrix properties. This trend is again verified in practice, although more variation between analytical and experimental results is observed in the examination of E_2 and G_{12} than for E_1 and v_{12}.

In order to attempt to obtain a closer agreement between analysis and experiment, a number of alternative approaches for the prediction of composite properties have been investigated. Some of these incorporate refinements to the approach previously described, perhaps the most successful and well-known being the procedure leading to the Halpin–Tsai equations,[1,3-5] while others are determined from the more advanced theory of elasticity.

Many of these various methods produce results which are more complex and unwieldy than the simple expressions obtained in this chapter, while some, although differing in the expressions obtained for E_2 and G_{12}, do in fact yield the same expressions for E_1 and v_{12} as are given by eqns. (5.6) and (5.11).

Thus it is considered that at the present stage of micromechanical development the prediction of the elastic constants of a composite lamina may be reasonably determined from an elementary strength of materials approach, although the reader is encouraged to become acquainted with the various alternative procedures.[1,3]

The determination of the elastic properties of laminae reinforced in two orthogonal directions with continuous reinforcement may be obtained from a combination of the results already derived.

5.3.2 Stress–Strain Relationships in a Discontinuous Fibre Lamina

Discontinuous fibre reinforcement is normally used in the form of random chopped strand mat which produces a lamina possessing isotropic properties. Due to this random nature of both direction and length of the individual fibres, an analytical determination of the properties of a lamina so reinforced becomes far more intractable than that for a continuous fibre composite.

However, to obtain an insight into the behaviour of a discontinuous fibre lamina, consider an element consisting of a single fibre of finite length l and circular cross-section of diameter d within a resin matrix as shown in Fig. 5.10. When the element is strained in the longitudinal direction it is seen that, due to the bond between the stiff fibre and the relatively flexible matrix, distortion of the matrix takes place in the vicinity of the ends of the fibre.

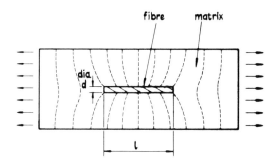

FIG. 5.10. Strain distribution in resin matrix around a fibre of finite length.

The predominant effect of this distortion is to produce shearing stresses along the fibre–matrix interface for some distance from the ends of the fibre; these stresses may be so high that some inelastic behaviour of the matrix may occur. The presence of these shearing stresses implies that the axial stress, σ_f, in the fibre decreases in these regions and, assuming an absence of strength between the ends of the fibre and the matrix, effectively becomes zero at the ends.[1,2,6-8]

To obtain an estimate of the length of the fibre over which the axial stress is reduced, assume that in the distorted region of the fibre–matrix boundary the matrix behaves perfectly plastically, i.e. the shearing stress along the fibre is constant and equal to the shearing strength of the matrix, τ_y. Then considering the stresses acting on an element of the fibre of length δ_x, as shown in Fig. 5.11, equilibrium considerations give

$$\tau_y(\pi d \delta x) = \delta \sigma_f \left(\pi \frac{d^2}{4} \right) \qquad (5.26)$$

or as δx tends to zero

$$\frac{d\sigma_f}{dx} = \frac{4\tau_y}{d} \tag{5.27}$$

Since $4\tau_y/d$ is a constant, σ_f thus varies linearly along the length, reaching its maximum value $\sigma_{f_{max}}$ at a distance along the fibre where the distortion is deemed to cease.

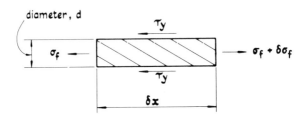

FIG. 5.11. Length of fibre subjected to interfacial shearing stress and corresponding axial stress.

Now from eqn. (5.27), it is seen that this distance along the fibre is independent of the length of the fibre itself, so for very short fibres, the value of $\sigma_{f_{max}}$ may never be reached. There is therefore a minimum length of fibre within which $\sigma_{f_{max}}$ will just be attained, and this length is known as the *critical length*, l_c.

This concept is made clear in Fig. 5.12, which shows this idealised axial stress distribution over three definitive lengths of fibre.

Thus eqn. (5.27) may be rewritten as

$$\frac{\sigma_{f_{max}}}{l_c/2} = \frac{4\tau_y}{d}$$

or

$$\frac{l_c}{d} = \sigma_{f_{max}} 2\tau_y \tag{5.28}$$

where l_c/d is known as the *critical aspect ratio*, i.e. the minimum length/diameter ratio of the fibre to ensure that $\sigma_{f_{max}}$ is attained.

These considerations imply that the axial force carried by a fibre of finite length is less than that sustained in a continuous fibre, and that the total force may be expressed in terms of an average tensile stress in the fibre, this average stress approaching $\sigma_{f_{max}}$ as the length of the fibre increases.

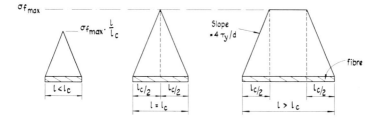

FIG. 5.12. Idealised axial stress distribution in fibres of various lengths.

The value of the average stress is obtained by equating the area of the appropriate stress distribution diagram given in Fig. 5.12 to the area of an equivalent rectangle. Thus for fibres with lengths not less than l_c, the average axial stress, $\sigma_{f_{av}}$, may be deduced from the relationship,

$$\sigma_{f\,max}(l-l_c)+\frac{\sigma_{f\,max}}{2}l_c=\sigma_{f_{av}}l \qquad (5.29)$$

which gives

$$\sigma_{f_{av}}=\sigma_{f\,max}\left(1-\frac{l_c}{2l}\right) \qquad (5.30)$$

The similar result for fibres whose length is less than l_c can be obtained using similar triangles as

$$\sigma_{f_{av}}=\frac{\sigma_{f\,max}l}{2l_c} \qquad (5.31)$$

Of course, it must be realised that these results can only be approximations, not least because the assumption of constant shearing stress of value τ_y at the fibre–matrix interface may not be totally valid, and elastic behaviour to a greater or lesser extent may be present. However in practice the stiffness of discontinuous fibre composites is found to be less than that of the equivalent continuously reinforced materials and therefore the simple derivations obtained are considered useful in the prediction of the elastic properties of such laminae.

If the stresses in the fibres are considered to be given by eqns. (5.30) and (5.31), the equation for the prediction of the longitudinal Young's modulus, E_1, for an element reinforced with unidirectional discontinuous fibres may be obtained by use of $\sigma_{f_{av}}$ instead of σ_f in eqn. (5.2), the equation used for the determination of E_1 in a continuously reinforced lamina. This is equivalent to reducing the value of Young's modulus for the fibres, E_f, in eqn. (5.5) and hence reducing the value of E_1 for the composite.

As has been mentioned previously, the use of random discontinuous fibres results in isotropic laminae of chopped strand mat form. Thus as an approximation to the single value of Young's modulus for such a lamina, one may consider half the fibres to act in one direction and half to act in the orthogonal direction, calculate the average length of fibre so that $\sigma_{f_{av}}$ may be determined, and so predict the modulus of elasticity from the modification of eqn. (5.6). The small additional effect given by eqns. (5.17) may be neglected. Poisson's ratio may be considered to be given by eqn. (5.11), and since the lamina is isotropic, the modulus of rigidity, G, follows from these two properties.

Further refinements have been made by various investigators to the descriptions given here, these including the effect of load transfer at the fibre ends and more realistic stress distributions along the fibre length,[9-11] and some recent theoretical work suggests that in the case of chopped strand mat only three-eights of the fibres should be considered to be effective in any one direction.[12]

Although the results presented here may be used to acquire some knowledge of the material properties, it must be realised that significant assumptions have been made to reduce the vast complexities inherent in the real material, and hence although these results may be used as a guide as to the properties, experimental confirmation is considered an essential ingredient in the determination of the elastic constants of such randomly reinforced materials.

5.4 THE STRENGTH CHARACTERISTICS OF GLASS-REINFORCED LAMINAE

The prediction of the ultimate strength of fibre-reinforced laminae is even more complex than the estimation of the elastic constants, and has had at present only limited success. The reasons for this become quite clear when one considers the additional variables involved over and above those already examined in the prediction of the elastic constants.

Firstly, failure of the composite may be due to either failure of the fibres or failure of the matrix, and secondly, because of the possibility of inelastic deformation occuring, usually in the matrix, there may be some similar behaviour in the composite prior to actual rupture, and so it may be deemed that failure has occurred when such behaviour becomes apparent. Also, in a fibre failure mode not all the fibres may fail simultaneously, a phenomenon that again may radically alter the composite prior to rupture; furthermore, at high stress degeneration of the fibre–resin bond may occur, so that the composite nature of the material is eroded.

However, simple and approximate analytical models, similar to those for the prediction of the elastic constants, can be developed which will give a guide as to the strength of composite laminae: some of these will now be briefly described.

5.4.1 Strengths of a Continuous Unidirectional Fibre Lamina

The micromechanical investigation of the strength of laminae continuously reinforced in one direction has had most success in the prediction of the ultimate tensile strength, X_t, in the direction of the fibres, and this topic has received most attention. Investigations into the compressive strength, X_c, in the direction of the fibres has tended to produce more complex analyses, and in general theoretical predictions have exceeded results obtained in practice.[2] Less work appears to have been carried out on the determination of the ultimate shearing strength, S, but it may be expected that this property, together with the tensile and compressive strengths in the directions normal to the fibres, Y_t and Y_c respectively, will bear a similarity to the corresponding matrix strength or to the matrix–fibre interfacial strength, whichever is less.

In this section two topics will be addressed, namely the prediction of the ultimate tensile and compressive strengths, X_t and X_c, in a continuously reinforced unidirectional lamina.

5.4.1.1 Ultimate Tensile Strength

During the application of an increasing tensile strain to a unidirectionally reinforced lamina in the fibre direction, i.e. the 1-direction, the deformation of the composite may be described approximately by a number of stages.

At relatively low stresses, the fibres and matrix deform elastically; then as the stress increases, the matrix, if it possesses any ductility, may begin to exhibit some inelastic deformation. Assuming that no such inelastic deformation occurs in the glass fibres, further increase of stress will initiate failure, usually by progressive fracture of the brittle fibres. In this qualitative description of lamina behaviour it is assumed that the fibre–resin interface remains intact throughout.

Although in practice the individual fibres vary in strength, the assumption will be made that they are all of equal strength, and also that they are more brittle than the matrix. During the deformation process, the strains in the fibres and matrix are the same, and hence

$$\varepsilon_1 = \frac{\sigma_f}{E_f} = \frac{\sigma_m}{E_m} \tag{5.32}$$

whilst from eqn. (5.4), used for the estimation of E_1,

$$\sigma_1 = \sigma_f V_f + \sigma_m V_m \tag{5.33}$$

Now since the fibres are more brittle than the matrix, i.e. the ultimate strain in the fibres is less than that in the matrix, the fibres will be the first component to fail. If the stress in the composite after failure of the fibres is greater than the ultimate strength of the matrix, then failure of the composite occurs at fracture of the fibres and eqn. (5.33) may be rewritten as

$$X_t = \sigma_{fu} V_f + \sigma_{mf} V_m \tag{5.34}$$

where X_t and σ_{fu} are the tensile failure stresses in the composite and fibres respectively, and σ_{mf} is the stress in the matrix at incipient failure of the fibres.

Thus if the fibres are to have any effect in increasing the strength of the composite above that of the matrix alone, i.e. above σ_{mu}, a minimum fibre volume fraction is required. This critical value, $V_{f_{crit}}$, can be obtained by rewriting eqn. (5.34) so that X_t is equal to σ_{mu}, and replacing V_f by $V_{f_{crit}}$ and V_m by $(1 - V_{f_{crit}})$:

$$\sigma_{mu} = \sigma_{fu} V_{f_{crit}} + \sigma_{mf}(1 - V_{f_{crit}}) \tag{5.35}$$

from which

$$V_{f_{crit}} = \frac{\sigma_{mu} - \sigma_{mf}}{\sigma_{fu} - \sigma_{mf}} \tag{5.36}$$

When the volume of fibres is very low, the stress in the composite after failure of the fibres may be lower than the ultimate strength of the matrix. In this situation, failure of the fibres does not imply failure of the composite, which does not reach the ultimate state until the matrix reaches its failure stress.

In this case the ultimate strength of the composite will be described by

$$X_t = \sigma_{mu} V_m = \sigma_{mu}(1 - V_f) \tag{5.37}$$

By combining eqns. (5.34) and (5.37) the manner in which the ultimate tensile strength of the composite depends on V_f can be described. This is shown graphically in Fig. 5.13, where it is noted that there is a value of the fibre volume fraction, $V_{f_{min}}$, which produces the minimum composite strength and simultaneous fibre and matrix failure.

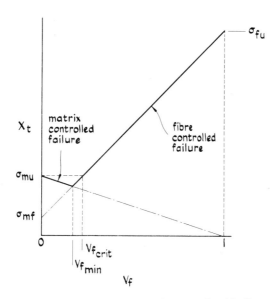

FIG. 5.13. Variation of ultimate tensile strength with fibre content.

The observations given above are based on the assumption that the fibres are all of equal strength and that they fracture simultaneously. In practice of course the fibres exhibit variations in strength so such idealised behaviour does not occur. Analyses based on a strength distribution in the fibres have been attempted,[13] but they are more complex and, although they are useful in obtaining a greater understanding of composite failure, they do not produce results which differ widely from the simpler estimates; therefore the grounds for their general adoption in the design of composites at present appear doubtful.

5.4.1.2 ULTIMATE COMPRESSIVE STRENGTH
When a unidirectional glass fibre-reinforced composite lamina is subjected to compression in the direction of the reinforcement, the failure load appears to be controlled by buckling of the fibres. To obtain an

estimate of the ultimate load in compression, therefore, it is assumed that each fibre acts independently of its neighbours and behaves as a column supported on an elastic foundation, the foundation in this case being the resin matrix.[14]

The two basic modes of buckling that can occur in this simplified model are shown in Fig. 5.14. At the ultimate load, adjacent fibres will buckle with equal wavelengths either

(a) out of phase with each other, predominantly producing extensional strains in the matrix, this deformation pattern being known as 'extensional' mode, or
(b) in phase with each other, producing shearing strains in the matrix, and hence deforming in 'shear' mode.

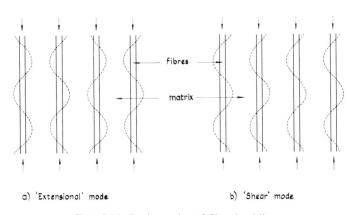

FIG. 5.14. Basic modes of fibre buckling.

The determination of the buckling stress for each mode is effected by equating the work done by the compressive fibre forces to the strain energy stored in the matrix and the fibres in deforming to the buckled state. The shapes of the buckled fibres are considered to be of sinusoidal form, the wavelength being determined as that which produces the lowest, or critical, buckling load.[1,3,6]

Applying this process to the two deformation modes proposed produces the following results for the ultimate compressive strengths, X_c, and critical strains, ε_{x_c}, of the composite:

For 'extensional' mode:

$$X_c = 2V_f \left[\frac{V_f E_m E_f}{3(1-V_f)} \right]^{1/2} \quad (5.38a)$$

$$\varepsilon_{x_c} = 2 \left[\frac{V_f}{3(1-V_f)} \cdot \frac{E_m}{E_f} \right]^{1/2} \quad (5.38b)$$

For 'shear' mode:

$$X_c = \frac{G_m}{1-V_f} \quad (5.39a)$$

$$\varepsilon_{x_c} = \frac{1}{V_f(1-V_f)} \left(\frac{G_m}{E_f} \right) \quad (5.39b)$$

where $1-V_f = V_m$.

For values of V_f below about 0·2 the 'extensional' mode is critical while for values of V_f greater than this, the 'shear' mode describes the ultimate behaviour of the composite. Figure 5.15 depicts this behaviour for a

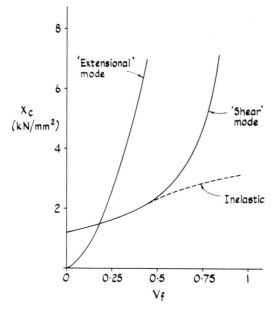

FIG. 5.15. Variation of ultimate compressive strength with fibre content. Material properties: $E_f = 72·4 \text{ kN/mm}^2$, $E_m = 3·22 \text{ kN/mm}^2$, $G_m = 1·19 \text{ kN/mm}^2$.

typical glass fibre/polyester resin lamina, and also indicates that as V_f is increased, the value of X_c obtained from the analysis becomes so high that in practice the matrix would begin to deform inelastically. This would effectively reduce the value of X_c and thus the 'shear' mode curve would tend to follow the lower line labelled 'inelastic'. That this is nearer to reality is borne out in practice, although experimental values of X_c tend to be lower than those predicted by the analysis described here.

Corresponding curves for estimating the relationship between the critical strain and V_f are shown in Fig. 5.16 where again it is seen that the 'shear' mode is dominant except for low values of V_f.

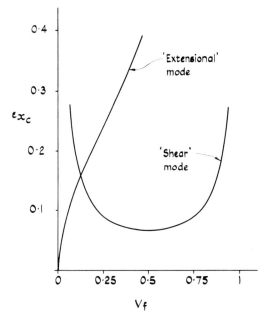

FIG. 5.16. Variation of critical compressive strain with fibre content. Material properties: $E_f = 72\cdot4 \text{ kN/mm}^2$, $E_m = 3\cdot22 \text{ kN/mm}^2$, $G_m = 1\cdot19 \text{ kN/mm}^2$.

5.4.2 Strengths of a Discontinuous Fibre Lamina

The nature of this form of composite is of course far more variable than that of a continuously reinforced material and hence any strength predictions applied to it must be considered in this light.

The tensile strength of a discontinuous fibre lamina may be predicted by firstly considering the case in which all the fibres lie in one direction.

Here the tensile strength may be estimated by means of equations similar to those used to predict tensile failure in a continuously reinforced lamina. If the critical length, l_c, of a fibre that will ensure that the ultimate tensile stress is attained is given from eqn. (5.28) as

$$l_c = \frac{\sigma_{fu}d}{2\tau_y}$$

it follows that the ultimate tensile stress in a discontinuous fibre will only be reached if the fibre length, l, is equal to or greater than this critical length. In this case the average tensile strength, $\sigma_{fu_{av}}$, of a fibre is obtained using eqn. (5.30):

$$\sigma_{fu_{av}} = \sigma_{fu}\left(1 - \frac{l_c}{2l}\right)$$

and the ultimate tensile strength of the lamina becomes, from eqn. (5.34),

$$X_t = \sigma_{fu_{av}} V_f + \sigma_{mf} V_m$$

$$= \sigma_{fu}\left(1 - \frac{l_c}{2l}\right)V_f + \sigma_{mf} V_m$$

$$= \sigma_{fu}\left(1 - \frac{\sigma_{fu}d}{4l\tau_y}\right)V_f + \sigma_{mf} V_m \qquad (5.40)$$

assuming that the critical value of the fibre volume fraction, now given by

$$V_{f_{crit}} = \frac{\sigma_{mu} - \sigma_{mf}}{\sigma_{fu_{av}} - \sigma_{mf}}$$

is not exceeded.

If the fibre volume fraction is less than this critical value, the ultimate matrix tensile strength determines the ultimate tensile strength of the lamina, which is then given by eqn. (5.37) as

$$X_t = \sigma_{mu} V_m$$

If the length of the fibre is less than the critical length, the ultimate fibre tensile stress can never be reached; the maximum stress in the fibre is then obtained from eqn. (5.31) as

$$\sigma_{f_{av}} = \frac{\sigma_{fu}l}{2l_c} \qquad (5.41)$$

From eqn. (5.27), σ_{fu} may be expressed as

$$\sigma_{fu} = \frac{4\tau_y}{d} \times \frac{l_c}{2} = \frac{2\tau_y l_c}{d} \tag{5.42}$$

and hence eqn. (5.41) may be written

$$\sigma_{f_{av}} = \frac{\tau_y l}{d} \tag{5.43}$$

Therefore in this situation, failure of the matrix defines failure of the composite, the ultimate tensile strength of which is

$$X_t = \frac{\tau_y l}{d} V_f + \sigma_{mu} V_m \tag{5.44}$$

In the case where the discontinuous fibres are randomly dispersed, producing an isotropic lamina, a tentative prediction of the ultimate tensile strength may be made by simply considering half the fibre volume fraction to be effective in any one direction. Alternative methods of estimating the strengths of such composites have been investigated[1] although these are generally rather more complex.

Micromechanical analyses for the compressive and shearing strengths of discontinuously reinforced laminae have not been developed to any extent and such properties, if required, should be furnished by experiment. It may be expected that the compressive strength will be lower than that for an equivalent continuously reinforced composite and that the shearing strength will be substantially matrix-dependent.

REFERENCES

1. AGARWAL, B. D. and BROUTMAN, L. J., *Analysis and Performance of Fiber Composites*, 1980, John Wiley, New York.
2. BROUTMAN, L. J. and KROCK, R. H., *Modern Composite Materials*, 1967, Addison-Wesley Publishing, Reading, Mass.
3. JONES, R. M., *Mechanics of Composite Materials*, 1975, McGraw-Hill, New York.
4. HALPIN, J. C. and THOMAS, R. L., Ribbon reinforcement of composites, *J. Composite Materials*, **2**, Oct. 1968, 488.
5. HALPIN, J. C. and TSAI, S. W., *Effects of Environmental Factors on Composite Materials*, AFML-TR 67-423, June 1969.
6. HOLISTER, G. S. and THOMAS, C., *Fibre Reinforced Materials*, 1966, Elsevier, Amsterdam.

7. HOLLAWAY, L., *Glass Reinforced Plastics in Construction*, 1978, Surrey University Press, Glasgow.
8. VINSON, J. R. and CHOU, T. W., *Composite Materials and their Use in Structures*, 1975, Applied Science Publishers, London.
9. CHON, C. T. and SUN, C. T., Stress distributions along a short fibre in fibre reinforced plastics, *J. Materials Science*, **15**, 1980, 931.
10. CHANG, D. C. and WENG, G. J., Elastic moduli of randomly orientated chopped-fibre composites with filled resin, *J. Materials Science*, **14**, 1979, 2183.
11. FUKUDA, H. and CHOU, T. W., Stiffness and strength of short fibre composites as affected by cracks and plasticity, *Fibre Science and Technology*, **15**, 1981, 243.
12. PHAN-THIEN, N. and HUILGOL, R. R., A micromechanic theory of chopped fibre-reinforced materials, *Fibre Science and Technology*, **13**, 1980, 423.
13. ROSEN, B. W., Tensile failure of fibrous composites, *AIAA Journal*, Nov. 1964, 1985.
14. TIMOSHENKO, S. P. and GERE, J. M., *Theory of Elastic Stability*, 2nd edn, 1961, McGraw-Hill, New York.

CHAPTER 6

The Behaviour of Glass Fibre-reinforced Laminates

6.1 INTRODUCTION

In the developments of the previous two chapters attention has been focused on attaining a knowledge of the behaviour of a single lamina of a glass-reinforced composite, this being the basic module in fibre-reinforced materials.

In practical applications of glass-reinforced plastics, the structural elements are not however single laminae, but rather several laminae bonded and stacked together to form a laminate of the thickness necessary for the required application. If each lamina is identical and isotropic in character, as would be the case if uniform layers of chopped strand mat reinforcement were used, it follows that the resulting laminate will also possess similar in-plane isotropic properties, while if the separate laminae possess orthotropy then the laminate can be designed to resist loads in several directions by suitably varying the orientation of the principal directions of each constituent lamina. In this latter case, because of the different orientations of the various laminae, the resulting laminate may lack any obvious principal directions.

In the construction of a typical laminate from a number of individual laminae (Fig. 6.1), one very important parameter is introduced that does not exist in single laminae, namely the strength of the laminate along planes between the individual laminae where little or no reinforcement exists. The importance of this parameter is evident when it is realised that, in general, any flexural action of a laminate will induce horizontal shearing stresses, and that hence the allowable magnitude of these will be highly dependent upon the shearing strength of the resin matrix.[1-4] Since this is relatively low, the interlaminar shearing stresses are a principal cause of delamination and hence are of paramount importance in the design of structural laminates.[5]

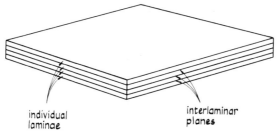

FIG. 6.1. Laminate construction.

It will be the aim of this chapter to examine the *macromechanical* behaviour of such laminates from the viewpoints of both stiffness and strength, making use of the results previously developed.

6.2 STIFFNESS CHARACTERISTICS OF LAMINATED COMPOSITES

The process of lamination satisfies two structural requirements. Firstly it produces a thickness of material sufficient to sustain high planar loads in various directions, and secondly, due to the resulting thickness attained, it enables lateral loading to be carried by the composite in bending.

In the descriptions of single laminae, only the planar effects are considered; negligible resistance to bending is encountered since the thickness is very small. Furthermore the material is considered uniform across the thickness of the lamina. As the thickness increases, the composite inherits increased resistance to flexure, while if the thickness is produced from several laminae whose properties in any given direction are different, then across the resulting thickness the material may be of a non-uniform nature.

In certain lamina stacking sequences, this resulting non-uniformity of the cross-sectional characteristics can produce behavioural complexities not incurred in materials with uniform cross-sectional material properties, one of the most striking of these being the interaction between planar and flexural action, from which bending can occur due to the application of planar forces.[1,3] This action is similar to that which occurs when a composite of two dissimilar metals is uniformly heated.

Although the purpose of this section is to develop the theory of generally laminated plates, the fundamental ideas are possibly more easily comprehended by firstly considering laminated beams, and developing the extension to plate theory from the results thus obtained.

6.2.1 Behaviour of Laminated Beams

Although structural laminates possess the characteristics of plates, there are many applications in which they may be considered as beams, the simplest being when a single plate spans in one direction only.

To develop the theory of laminated beams, consideration will be given to the effect of axial and flexural forces on three beams of rectangular section and of various material characteristics. Firstly the elementary case of the isotropic beam will be stated, and this will be followed by an examination of beams with symmetrical and non-symmetrical variation in cross-sectional material properties.

6.2.1.1 Isotropic Beam

Although the simplest case to consider, the results obtained for all other cases of laminated beams and plates may be considered to be developments of the results for an isotropic beam.

Consider the portion of length δx of an isotropic beam shown in Fig. 6.2a. Making the initial assumption that plane sections of the beam perpendicular to the middle line remain plane and perpendicular after deformation, i.e. assuming that the effect of any shearing strain on the strains due to flexure is negligible,[6,7] a justifiable assumption unless the beam is very deep, under any general system of axial and flexural loading the beam will deform so that any section BB will be displaced relative to an adjacent section AA. Reasonably neglecting any deformation due to shear, this displacement may be decomposed into a translation δu_0 due to axial effects (Fig. 6.2b) and a rotation $\delta \theta$ due to flexural effects (Fig. 6.2c). If these displacements are small, or more strictly tend to zero, then the total displacement in the x-direction of section BB relative to section AA at a point distance z from the middle line is

$$\delta u = \delta u_0 - z \delta \theta \tag{6.1}$$

when δu_0 and $z \delta \theta$ are the displacements due to axial and flexural effects respectively.

Hence the strain in the line distance z from the middle line becomes

$$\varepsilon = \frac{\delta u}{\delta x} = \frac{\delta u_0}{\delta x} - z \frac{\delta \theta}{\delta x} \tag{6.2a}$$

or in the limit

$$\varepsilon = \frac{du}{dx} = \frac{du_0}{dx} - z \frac{d\theta}{dx} \tag{6.2b}$$

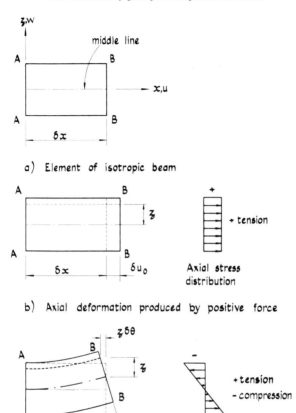

FIG. 6.2. Deformations and stress distributions in isotropic beam.

Now the rate of change of rotation $d\theta/dx$ can be expressed in terms of the lateral deflection of the beam, w, as

$$\frac{d\theta}{dx} = \frac{d}{dx}\left(\frac{dw}{dx}\right) = \frac{d^2w}{dx^2} \qquad (6.3)$$

and thus eqn. (6.2b) can be rewritten as

$$\varepsilon = \frac{du}{dx} = \frac{du_0}{dx} - z\frac{d^2w}{dx^2} \qquad (6.4)$$

The longitudinal stress, σ, at distance z from the middle line may now be simply expressed as

$$\sigma = E\varepsilon = E\left(\frac{du_0}{dx} - z\frac{d^2w}{dx^2}\right) \tag{6.5}$$

Thus since the strain distribution is linear, so also is that due to the stress.

From eqn. (6.5) the resultant axial forces and bending moments causing the displacements can now be developed. From Fig. 6.3, it may be seen that by appropriate integration over the thickness t of the beam, the resultant axial force, N, and bending moment, M, per unit width of the beam are respectively given by

$$N = \int_{-t/2}^{t/2} \sigma \, dz \tag{6.6a}$$

and

$$M = -\int_{-t/2}^{t/2} \sigma z \, dz \tag{6.6b}$$

the negative sign being necessary by virtue of the convention for positive bending moment defined in Fig. 6.2c.

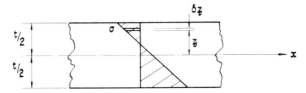

FIG. 6.3. Total stress distribution in isotropic beam.

On substituting eqn. (6.5) into eqns. (6.6a) and (6.6b) the following results are obtained:

$$\begin{aligned} N &= E\int_{-t/2}^{t/2} \left(\frac{du_0}{dx} - z\frac{d^2w}{dx^2}\right) dz \\ &= E\frac{du_0}{dx}[z]_{-t/2}^{t/2} - E\frac{d^2w}{dx^2}\left[\frac{z^2}{2}\right]_{-t/2}^{t/2} \\ &= Et\frac{du_0}{dx} \end{aligned} \tag{6.7}$$

and

$$M = -E \int_{-t/2}^{t/2} \left(z \frac{du_0}{dx} - z^2 \frac{d^2w}{dx^2} \right) dz$$

$$= -E \frac{du_0}{dx} \left[\frac{z^2}{2} \right]_{-t/2}^{t/2} + E \frac{d^2w}{dx^2} \left[\frac{z^3}{3} \right]_{-t/2}^{t/2}$$

$$= E \frac{t^3}{12} \frac{d^2w}{dx^2} \tag{6.8}$$

Hence it can be seen that eqns. (6.7) and (6.8) represent the standard equations for axial and flexural effects in a rectangular beam of unit width.

This analysis, as has been previously stated, is the basis for the development of the general laminate theory for beams and plates.

6.2.1.2 Three-layered Symmetrically Laminated Beam

Consider now the behaviour of a beam composed of three layers, the outer two of which have identical thicknesses and properties while the central one possesses different geometrical and material characteristics. A portion of such a beam is described in Fig. 6.4a, in which material symmetry exists about the middle line.

Sections originally plane and perpendicular to the middle line are again assumed to remain so after deformation, so that the same results for strain that were developed for the isotropic beam are maintained. Thus eqns. (6.1), (6.2), (6.3) and (6.4) also apply for this beam.

What does not follow, however, is the result for stress given by eqn. (6.5). Due to the variation in Young's modulus between the various layers, the stress diagram will exhibit a stepwise linear character rather than the continuous one observed previously. Hence eqn. (6.5) must be replaced by two separate equations, namely

$$\sigma = E_a \left(\frac{du_0}{dx} - z \frac{d^2w}{dx^2} \right) \tag{6.9a}$$

for values of z over the outer layers, where Young's modulus is E_a, and

$$\sigma = E_b \left(\frac{du_0}{dx} - z \frac{d^2w}{dx^2} \right) \tag{6.9b}$$

for values of z over the middle layer where Young's modulus is E_b.

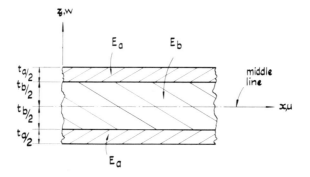

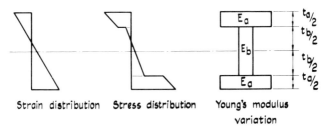

a) Portion of three-layered beam

b) Variation of stress, strain, and elastic moduli

FIG. 6.4. Symmetrically laminated beam.

The variation of strain, stress and elastic moduli is shown in Fig. 6.4b.

The resultant axial force and bending moment per unit width may now be obtained as before, except that now they must be initially expressed as the sum of the three constituent portions, thus:

$$N = \int_{-(t_a+t_b)/2}^{-t_b/2} \sigma \, dz + \int_{-t_b/2}^{t_b/2} \sigma \, dz + \int_{t_b/2}^{(t_a+t_b)/2} \sigma \, dz \quad (6.10a)$$

and

$$M = -\int_{-(t_a+t_b)/2}^{-t_b/2} \sigma z \, dz - \int_{-t_b/2}^{t_b/2} \sigma z \, dz - \int_{t_b/2}^{(t_a+t_b)/2} \sigma z \, dz \quad (6.10b)$$

whereupon substituting the relevant expressions for stress in terms of the strain components gives

$$N = E_a \int_{-(t_a+t_b)/2}^{-t_b/2} \left(\frac{du_0}{dx} - z\frac{d^2w}{dx^2}\right) dz + E_b \int_{-t_b/2}^{t_b/2} \left(\frac{du_0}{dx} - z\frac{d^2w}{dx^2}\right) dz$$

$$+ E_a \int_{t_b/2}^{(t_a+t_b)/2} \left(\frac{du_0}{dx} - z\frac{d^2w}{dx^2}\right) dz = (E_a t_a + E_b t_b)\frac{du_0}{dx} \quad (6.11)$$

and

$$M = -E_a \int_{-(t_a+t_b)/2}^{-t_b/2} \left(z\frac{du_0}{dx} - z^2\frac{d^2w}{dx^2}\right) dz - E_b \int_{-t_b/2}^{t_b/2} \left(z\frac{du_0}{dx} - z^2\frac{d^2w}{dx^2}\right) dz$$

$$- E_a \int_{t_b/2}^{(t_a+t_b)/2} \left(z\frac{du_0}{dx} - z^2\frac{d^2w}{dx^2}\right) dz = \frac{1}{12}(E_a[\{t_a+t_b\}^3 - t_b^3] + E_b t_b^3)\frac{d^2w}{dx^2} \quad (6.12)$$

It can be seen that the stiffness expression in eqn. (6.11) is simply a 'rule of mixtures' result, each term representing the stiffness of the respective forms of layer, while that in eqn. (6.12) is obtained from the separate second moments of area for the individual sections. It may be further noted that if E_b were zero, then the results would simply reduce to the properties of a symmetrical I-section where only the effect of the flanges are considered significant.

6.2.1.3 TWO-LAYERED NON-SYMMETRICALLY LAMINATED BEAM

Finally consider the behaviour of a beam laminated unsymmetrically with respect to the middle line.

The simplest form of such a beam is one composed of only two layers, each of which has different material properties (Fig. 6.5a). Again making the previous assumption regarding plane and perpendicular sections, the strain, stress and Young's modulus variation over a section of the beam is shown in Fig. 6.5b. The expressions for the resultant axial force and bending moments may now be written similarly to the earlier ones:

$$N = E_a \int_{-(t_a+t_b)/2}^{(t_a-t_b)/2} \left(\frac{du_0}{dx} - z\frac{d^2w}{dx^2}\right) dz + E_b \int_{(t_a-t_b)/2}^{(t_a+t_b)/2} \left(\frac{du_0}{dx} - z\frac{d^2w}{dx^2}\right) dz$$

$$= (E_a t_a + E_b t_b)\frac{du_0}{dx} + (E_a - E_b)\frac{t_a t_b}{2}\frac{d^2w}{dx^2} \quad (6.13)$$

and

$$M = -E_a \int_{-(t_a+t_b)/2}^{(t_a-t_b)/2} \left(z\frac{du_0}{dx} - z^2\frac{d^2w}{dx^2}\right)dz - E_b \int_{(t_a-t_b)/2}^{(t_a+t_b)/2} \left(z\frac{du_0}{dx} - z^2\frac{d^2w}{dx^2}\right)dz$$

$$= (E_a - E_b)\frac{t_a t_b}{2}\frac{du_0}{dx} + \frac{1}{12}(E_a t_a[t_a^2 + 3t_b^2] + E_b t_b[3t_a^2 + t_b^2])\frac{d^2w}{dx^2} \quad (6.14)$$

The difference between these expressions and the corresponding ones for the symmetrical case is now plain. Whereas for a symmetrically laminated beam an axial force produces only axial displacement, and a bending moment only flexural deformation, in a non-symmetrically laminated beam the two effects are coupled so that the separate applications of axial force and bending moment produce both axial and flexural deformations. Similar, but naturally more complex, expressions may be deduced for non-symmetrically laminated beams possessing any number of laminae by simple extension of the analytical process.

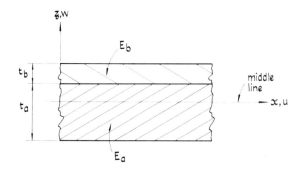

a) Portion of two-layered beam

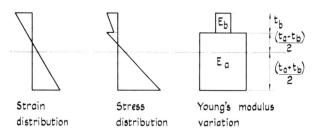

b) Variation of stress, strain, and elastic moduli

FIG. 6.5. Non-symmetrically laminated beam.

6.2.1.4 GENERAL EQUATIONS FOR LAMINATED BEAMS

The results obtained for symmetrically and non-symmetrically laminated beams (the isotropic case being regarded as a special case of symmetry), imply that the relationships for any laminated beam may be expressed in the form

$$N = A\frac{du_0}{dx} + B\frac{d^2w}{dx^2} \tag{6.15a}$$

$$M = B\frac{du_0}{dx} + D\frac{d^2w}{dx^2} \tag{6.15b}$$

or in matrix form

$$\begin{bmatrix} N \\ M \end{bmatrix} = \begin{bmatrix} A & B \\ B & D \end{bmatrix} \begin{bmatrix} \dfrac{du_0}{dx} \\ \dfrac{d^2w}{dx^2} \end{bmatrix} \tag{6.16}$$

where symmetry about the leading diagonal can be noted.

In the case of symmetrically laminated beams, the coefficients B disappear, these representing the coupling effect present in non-symmetrical lamination, while for an isotropic beam the coefficients A and D become simply the axial and flexural rigidities of a beam of thickness t and unit width, namely Et and $Et^3/12$ respectively.

6.2.2 Behaviour of Laminated Plates

The logical extension of one-dimensional beam theory to two-dimensional plate theory is performed by considering, instead of a linear element of length δx as shown in Fig. 6.2, a rectangular element whose sides are δx and δy. Such an element is depicted in Fig. 6.6. It is now observed that the middle line of the beam element must, in the case of the plate, be replaced by a middle plane. Then instead of examining the stress and strain in a line some distance z from a middle line, the in-plane behaviour of a lamina a representative distance z from the middle plane must be investigated.

It is possible for such a lamina to undergo normal deformation in the two directions, x and y, and also to suffer distortion in its plane due to shear. Again making the assumption that plane and perpendicular sections remain so after deformation, the displacements δu, δv in the x-

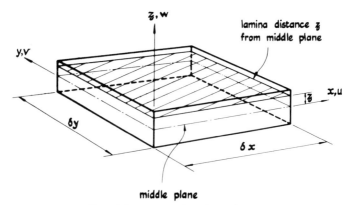

FIG. 6.6. Rectangular plate element.

and y-directions respectively may be written similarly to eqn. (6.1) as

$$\delta u = \delta u_0 - z \frac{\delta w}{\delta x} \qquad (6.17a)$$

and

$$\delta v = \delta v_0 - z \frac{\delta w}{\delta y} \qquad (6.17b)$$

and since the normal and shearing strains may be expressed as

$$\varepsilon_x = \frac{\partial u}{\partial x}$$

$$\varepsilon_y = \frac{\partial v}{\partial y}$$

$$\gamma_{xy} = \frac{\partial u}{\partial y} + \frac{\partial v}{\partial x} \qquad (6.18)$$

by invoking eqns. (6.17) these strains may be written as

$$\varepsilon_x = \frac{\partial u_0}{\partial x} - z \frac{\partial^2 w}{\partial x^2}$$

$$\varepsilon_y = \frac{\partial v_0}{\partial y} - z \frac{\partial^2 w}{\partial y^2}$$

$$\gamma_{xy} = \frac{\partial u_0}{\partial y} + \frac{\partial v_0}{\partial x} - 2z\frac{\partial^2 w}{\partial x \partial y} \quad (6.19)$$

It is seen that these equations are simply an extension of eqn. (6.4) and that partial differentiation must be implemented since u, v and w are in general functions of both x and y.

At this stage it is convenient to write eqns. (6.19) in a short matrix form.
Letting

$$[\varepsilon_0] = \begin{bmatrix} \dfrac{\partial u_0}{\partial x} \\ \dfrac{\partial v_0}{\partial y} \\ \dfrac{\partial u_0}{\partial y} + \dfrac{\partial v_0}{\partial x} \end{bmatrix} \quad (6.20)$$

and

$$[\chi] = \begin{bmatrix} \dfrac{\partial^2 w}{\partial x^2} \\ \dfrac{\partial^2 w}{\partial y^2} \\ 2\dfrac{\partial^2 w}{\partial x \partial y} \end{bmatrix} \quad (6.21)$$

then eqns. (6.19) may be written briefly as

$$\begin{bmatrix} \varepsilon_x \\ \varepsilon_y \\ \gamma_{xy} \end{bmatrix} = [\varepsilon] = [\varepsilon_0] - z[\chi] \quad (6.22)$$

thus preserving formal similarity with eqn. (6.4).

6.2.2.1 Isotropic Plate

As in the case of the beam, the simplest form of plate is one which displays material isotropy.

Analogously to the single normal stress–strain relationship for an isotropic beam given in eqn. (6.5), the three stresses in the plane of a lamina of the plate, namely σ_x, σ_y and τ_{xy}, may be written in terms of the corresponding strains ε_x, ε_y and γ_{xy} as

$$\begin{bmatrix} \sigma_x \\ \sigma_y \\ \tau_{xy} \end{bmatrix} = \begin{bmatrix} \dfrac{E}{1-v^2} & \dfrac{vE}{1-v^2} & \\ \dfrac{vE}{1-v^2} & \dfrac{E}{1-v^2} & \\ & & \dfrac{E}{2(1+v)} \end{bmatrix} \begin{bmatrix} \varepsilon_x \\ \varepsilon_y \\ \gamma_{xy} \end{bmatrix} \quad (6.23)$$

or more shortly

$$[\sigma] = [Q][\varepsilon]$$
$$= [Q]([\varepsilon_0] - z[\chi]) \quad (6.24)$$

From these equations the resultant forces and moments in the plate can now be determined in a manner analogous to that used in the beam. The difference between the forces in the beam and those in the plate is that, whereas in the beam the one stress develops in general one in-plane force, N, and one bending moment, M, in the plate the three stresses σ_x, σ_y and τ_{xy} produce three corresponding in-plane forces, namely two axial forces N_x, N_y, and one shearing force N_{xy}, and also three corresponding moments, consisting of two bending moments M_x, M_y, and one torque M_{xy}, these forces and moments being necessarily expressed per unit width of the plate since they will in general vary across its width. Because of the linearity of the strain distribution across a section the three stresses also show linearity and can be represented as shown in Fig. 6.7.

The resultant forces and moments may now be obtained by integration across the thickness t of the plate giving

$$N_x = \int_{-t/2}^{t/2} \sigma_x \, dz$$

$$N_y = \int_{-t/2}^{t/2} \sigma_y \, dz \quad (6.25a)$$

$$N_{xy} = \int_{-t/2}^{t/2} \tau_{xy} \, dz$$

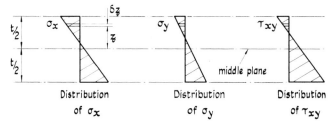

FIG. 6.7. Total normal and shearing stress distributions in isotropic plate.

and

$$M_x = -\int_{-t/2}^{t/2} \sigma_x z \, dz$$

$$M_y = -\int_{-t/2}^{t/2} \sigma_y z \, dz \quad (6.25b)$$

$$M_{xy} = -\int_{-t/2}^{t/2} \tau_{xy} z \, dz$$

the negative sign in the last three expressions occurring by virtue of the sign convention for flexure discussed previously with regard to the bending of beams (Section 6.2.1).

Hence on substituting eqns. (6.23) into eqns. (6.25a), and using eqns. (6.19), the values of the in-plane forces per unit width of the plate are given by:

$$N_x = \frac{E}{1-v^2} \int_{-t/2}^{t/2} \left(\frac{\partial u_0}{\partial x} - z \frac{\partial^2 w}{\partial x^2} \right) dz + \frac{vE}{1-v^2} \int_{-t/2}^{t/2} \left(\frac{\partial v_0}{\partial y} - z \frac{\partial^2 w}{\partial y^2} \right) dz$$

$$N_y = \frac{vE}{1-v^2} \int_{-t/2}^{t/2} \left(\frac{\partial u_0}{\partial x} - z \frac{\partial^2 w}{\partial x^2} \right) dz + \frac{E}{1-v^2} \int_{-t/2}^{t/2} \left(\frac{\partial v_0}{\partial y} - z \frac{\partial^2 w}{\partial y^2} \right) dz \quad (6.26)$$

$$N_{xy} = \frac{E}{2(1+v)} \int_{-t/2}^{t/2} \left(\frac{\partial u_0}{\partial y} + \frac{\partial v_0}{\partial x} - 2z \frac{\partial^2 w}{\partial x \partial y} \right) dz$$

which integrating become

$$N_x = \frac{Et}{1-v^2}\left(\frac{\partial u_0}{\partial x} + v\frac{\partial v_0}{\partial y}\right)$$

$$N_y = \frac{Et}{1-v^2}\left(\frac{\partial v_0}{\partial y} + v\frac{\partial u_0}{\partial x}\right) \quad (6.27)$$

$$N_{xy} = \frac{Et}{2(1+v)}\left(\frac{\partial u_0}{\partial x} + \frac{\partial v_0}{\partial y}\right)$$

Similarly, substituting eqn. (6.23) into eqn. (6.25b), and again making use of eqns. (6.19), gives the moments per unit width as

$$M_x = -\frac{E}{1-v^2}\int_{-t/2}^{t/2}\left(z\frac{\partial u_0}{\partial x} - z^2\frac{\partial^2 w}{\partial x^2}\right)dz - \frac{vE}{1-v^2}\int_{-t/2}^{t/2}\left(z\frac{\partial v_0}{\partial y} - z^2\frac{\partial^2 w}{\partial y^2}\right)dz$$

$$M_y = -\frac{vE}{1-v^2}\int_{-t/2}^{t/2}\left(z\frac{\partial u_0}{\partial x} - z^2\frac{\partial^2 w}{\partial x^2}\right)dz - \frac{E}{1-v^2}\int_{-t/2}^{t/2}\left(z\frac{\partial v_0}{\partial y} - z^2\frac{\partial^2 w}{\partial y^2}\right)dz$$

$$M_{xy} = -\frac{E}{2(1+v)}\int_{-t/2}^{t/2}\left(z\frac{\partial u_0}{\partial x} + z\frac{\partial v_0}{\partial y} - 2z^2\frac{\partial^2 w}{\partial x \partial y}\right)dz \quad (6.28)$$

thus giving[8]

$$M_x = \frac{Et^3}{12(1-v^2)}\left(\frac{\partial^2 w}{\partial x^2} + v\frac{\partial^2 w}{\partial y^2}\right)$$

$$M_y = \frac{Et^3}{12(1-v^2)}\left(\frac{\partial^2 w}{\partial y^2} + v\frac{\partial^2 w}{\partial x^2}\right) \quad (6.29)$$

$$M_{xy} = \frac{Et^3}{24(1+v)}\left(2\frac{\partial^2 w}{\partial x \partial y}\right) = \frac{Et^3}{12(1+v)}\frac{\partial^2 w}{\partial x \partial y}$$

The correlation between eqns. (6.27) and (6.29) for the plate and eqns. (6.7) and (6.8) for the beam cannot be missed, especially if v is zero for the material, as then the expressions for the axial forces and bending moments become identical.

Thus, with the sign convention utilised in the foregoing analysis, the forces and moments acting on an element of the plate are as shown in Fig. 6.8, where it is seen that because of the necessity of equilibrating complementary shearing stresses, complementary shearing forces and torques are automatically produced.

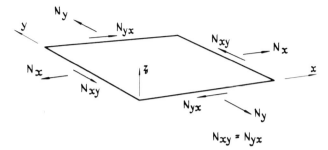

Positive in-plane forces

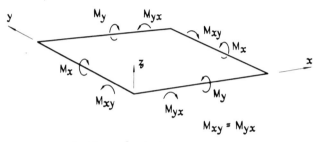

Positive out-of-plane moments

FIG. 6.8. In-plane forces and out-of plane moments acting on rectangular plate element.

6.2.2.2 Orthotropic Plate

The extension of isotropic plate theory to the investigation of the behaviour of a plate possessing orthotropy in the plane of the plate, whilst retaining uniform properties across the thickness, requires modifications to eqns. (6.23) to describe the more complex stress–strain relationships.

For the case when the stresses and strains are defined along the principal material axes, 1 and 2, of the plate, eqns. (6.23) may simply be modified to

$$\begin{bmatrix} \sigma_1 \\ \sigma_2 \\ \tau_{12} \end{bmatrix} = \begin{bmatrix} \dfrac{E_1}{1-v_{12}v_{21}} & \dfrac{v_{21}E_1}{1-v_{12}v_{21}} & \\ \dfrac{v_{12}E_2}{1-v_{12}v_{21}} & \dfrac{E_2}{1-v_{12}v_{21}} & \\ & & G_{12} \end{bmatrix} \begin{bmatrix} \varepsilon_1 \\ \varepsilon_2 \\ \gamma_{12} \end{bmatrix} \quad (6.30)$$

but when the stresses and strains are defined about any other set of orthogonal axes, x and y, coupling between normal and shearing effects takes place and the stress–strain relationships take the form

$$\begin{bmatrix} \sigma_x \\ \sigma_y \\ \tau_{xy} \end{bmatrix} = \begin{bmatrix} \bar{Q}_{11} & \bar{Q}_{12} & \bar{Q}_{13} \\ \bar{Q}_{12} & \bar{Q}_{22} & \bar{Q}_{23} \\ \bar{Q}_{13} & \bar{Q}_{23} & \bar{Q}_{33} \end{bmatrix} \begin{bmatrix} \varepsilon_x \\ \varepsilon_y \\ \gamma_{xy} \end{bmatrix} \qquad (6.31)$$

as described in Chapter 4, where the coefficients $\bar{Q}_{ij}$ are the functions of the material properties in the principal material directions and of the relative orientation between the two coordinate systems. Again the general stress–strain relationships can be expressed in the shortened form,

$$\begin{aligned} [\sigma] &= [\bar{Q}][\varepsilon] \\ &= [\bar{Q}]([\varepsilon_0] - z[\chi]) \end{aligned} \qquad (6.32)$$

Because of uniformity across the thickness, eqns. (6.25) still apply so the values of the forces and moments per unit width may be calculated as before. Thus

$$N_x = \bar{Q}_{11} \int_{-t/2}^{t/2} \left(\frac{\partial u_0}{\partial x} - z \frac{\partial^2 w}{\partial x^2} \right) dz + \bar{Q}_{12} \int_{-t/2}^{t/2} \left(\frac{\partial v_0}{\partial y} - z \frac{\partial^2 w}{\partial y^2} \right) dz$$

$$+ \bar{Q}_{13} \int_{-t/2}^{t/2} \left(\frac{\partial u_0}{\partial y} + \frac{\partial v_0}{\partial x} - 2z \frac{\partial^2 w}{\partial x \partial y} \right) dz \qquad (6.33)$$

N_y and N_{xy} may be similarly obtained by substitution of the relevant coefficients of $[\bar{Q}]$, while

$$M_x = -\bar{Q}_{11} \int_{-t/2}^{t/2} \left(z \frac{\partial u_0}{\partial x} - z^2 \frac{\partial^2 w}{\partial x^2} \right) dz - \bar{Q}_{12} \int_{-t/2}^{t/2} \left(z \frac{\partial v_0}{\partial y} - z^2 \frac{\partial^2 w}{\partial y^2} \right) dz$$

$$- \bar{Q}_{13} \int_{-t/2}^{t/2} \left(z \frac{\partial u_0}{\partial y} + z \frac{\partial v_0}{\partial x} - 2z^2 \frac{\partial^2 w}{\partial x \partial y} \right) dz \qquad (6.34)$$

with similar expressions obtainable for M_y and M_{xy}.

On carrying out the integrations, the forces per unit width become

$$N_x = \left(\bar{Q}_{11}\frac{\partial u_0}{\partial x} + \bar{Q}_{12}\frac{\partial v_0}{\partial y} + \bar{Q}_{13}\left[\frac{\partial u_0}{\partial y} + \frac{\partial v_0}{\partial x}\right]\right)t$$

$$N_y = \left(\bar{Q}_{12}\frac{\partial u_0}{\partial x} + \bar{Q}_{22}\frac{\partial v_0}{\partial y} + \bar{Q}_{23}\left[\frac{\partial u_0}{\partial y} + \frac{\partial v_0}{\partial x}\right]\right)t \quad (6.35)$$

$$N_{xy} = \left(\bar{Q}_{13}\frac{\partial u_0}{\partial x} + \bar{Q}_{23}\frac{\partial v_0}{\partial y} + \bar{Q}_{33}\left[\frac{\partial u_0}{\partial y} + \frac{\partial v_0}{\partial x}\right]\right)t$$

or, noting eqn. (6.20) and letting

$$[N] = \begin{bmatrix} N_x \\ N_y \\ N_{xy} \end{bmatrix}$$

then

$$[N] = [\bar{Q}]t[\varepsilon_0] = [A][\varepsilon_0] \quad (6.36)$$

while the moments per unit width are

$$M_x = \left(\bar{Q}_{11}\frac{\partial^2 w}{\partial x^2} + \bar{Q}_{12}\frac{\partial^2 w}{\partial y^2} + 2\bar{Q}_{13}\frac{\partial^2 w}{\partial x \partial y}\right)\frac{t^3}{12}$$

$$M_y = \left(\bar{Q}_{12}\frac{\partial^2 w}{\partial x^2} + \bar{Q}_{22}\frac{\partial^2 w}{\partial y^2} + 2\bar{Q}_{23}\frac{\partial^2 w}{\partial x \partial y}\right)\frac{t^3}{12} \quad (6.37)$$

$$M_{xy} = \left(\bar{Q}_{13}\frac{\partial^2 w}{\partial x^2} + \bar{Q}_{23}\frac{\partial^2 w}{\partial y^2} + 2\bar{Q}_{33}\frac{\partial^2 w}{\partial x \partial y}\right)\frac{t^3}{12}$$

or noting eqn. (6.21) and letting

$$[M] = \begin{bmatrix} M_x \\ M_y \\ M_{xy} \end{bmatrix}$$

then

$$[M] = [\bar{Q}]\frac{t^3}{12}[\chi] = [D][\chi] \quad (6.38)$$

Naturally these equations reduce to the isotropic case given by eqns. (6.27) and (6.29) when the values of Young's moduli E_1 and E_2 and Poisson's ratios v_{12} and v_{21} become equal, and thus eqns. (6.35) and (6.37) may be considered to represent any laminate, isotropic or orthotropic, providing that the material properties throughout the depth of the plate are uniform. Such plate laminates that have so far been discussed are in practice composed of a number of laminae of identical material properties and orientation.

Attention must now be focused on the development of the stiffness properties of a laminate composed of a number of laminae which possess different planar properties. From the investigation of laminated beams it is apparent that if symmetrical lamination occurs, then the in-plane and out-of-plane behaviour will be independent of each other, whereas if the plate is non-symmetrically laminated, coupling will occur between the two modes of behaviour. To illustrate the properties of each form of laminate, two examples will be considered, the first being a three-layered symmetric laminate and the second a two-layered non-symmetric laminate.

6.2.2.3 THREE-LAYERED SYMMETRIC LAMINATE

Consider a plate constructed from three generally orthotropic laminae, the top and bottom laminae being identical. An exploded view of such a plate is shown in Fig. 6.9.

If the stress–strain relationships of the top and bottom laminae, each of thickness $t_a/2$, are given by the equations

$$\begin{bmatrix} \sigma_x \\ \sigma_y \\ \tau_{xy} \end{bmatrix} = [\bar{Q}]_a \begin{bmatrix} \varepsilon_x \\ \varepsilon_y \\ \gamma_{xy} \end{bmatrix} \tag{6.39}$$

and those of the middle layer of thickness t_b by

$$\begin{bmatrix} \sigma_x \\ \sigma_y \\ \tau_{xy} \end{bmatrix} = [\bar{Q}]_b \begin{bmatrix} \varepsilon_x \\ \varepsilon_y \\ \gamma_{xy} \end{bmatrix} \tag{6.40}$$

then the forces and moments per unit width of the plate may be obtained by addition of the various effects of the separate laminae in a similar manner to that presented for a symmetrically laminated beam in eqns. (6.10).

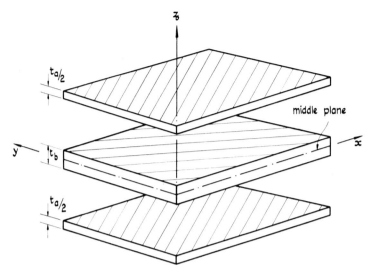

FIG. 6.9. Exploded view of a general three-layered symmetric laminate.

Thus the axial and shearing forces per unit width of the laminate are

$$N_x = \int_{-(t_a+t_b)/2}^{-t_b/2} \sigma_x \, dz + \int_{-t_b/2}^{t_b/2} \sigma_x \, dz + \int_{t_b/2}^{(t_a+t_b)/2} \sigma_x \, dz$$

$$N_y = \int_{-(t_a+t_b)/2}^{-t_b/2} \sigma_y \, dz + \int_{-t_b/2}^{t_b/2} \sigma_y \, dz + \int_{t_b/2}^{(t_a+t_b)/2} \sigma_y \, dz \qquad (6.41)$$

$$N_{xy} = \int_{-(t_a+t_b)/2}^{-t_b/2} \tau_{xy} \, dz + \int_{-t_b/2}^{t_b/2} \tau_{xy} \, dz + \int_{t_b/2}^{(t_a+t_b)/2} \tau_{xy} \, dz$$

The first and third term in each of these expressions represents the behaviour of the top and bottom layers respectively, while the second term represents the behaviour of the middle lamina.

On substituting eqns. (6.39) and (6.40), eqns. (6.41) may be written

$$[N] = [\bar{Q}]_a \int_{-(t_a+t_b)/2}^{-t_b/2} [\varepsilon] \, dz + [\bar{Q}]_b \int_{-t_b/2}^{t_b/2} [\varepsilon] \, dz$$

$$+ [\bar{Q}]_a \int_{t_b/2}^{(t_a+t_b)/2} [\varepsilon] \, dz \qquad (6.42)$$

where $[\varepsilon]$ may be further replaced by $([\varepsilon_0] - z[\chi])$.

On performing the integrations in the right-hand side expressions of eqns. (6.42) the result

$$[N] = ([\bar{Q}]_a t_a + [\bar{Q}]_b t_b)[\varepsilon_0] \tag{6.43}$$

is obtained, which is a simple 'rule of mixtures' equation.

Equations (6.42) and (6.43) may now be directly compared with eqn. (6.11). Indeed, eqn. (6.11) may be considered to be a special case of eqns. (6.42) and (6.43) where each of the matrices contains only one element. It is in mathematical situations such as this that the matrix formulations demonstrate their real power in being able to penetrate to the fundamental nature of seemingly complex problems.

It is now an elementary process to develop similar equations for the bending moments and torques per unit width of the plate. Analogously to eqns. (6.41), these may be expressed as

$$M_x = -\int_{-(t_a+t_b)/2}^{-t_b/2} \sigma_x z \, dz - \int_{-t_b/2}^{t_b/2} \sigma_x z \, dz - \int_{t_b/2}^{(t_a+t_b)/2} \sigma_x z \, dz$$

$$M_y = -\int_{-(t_a+t_b)/2}^{-t_b/2} \sigma_y z \, dz - \int_{-t_b/2}^{t_b/2} \sigma_y z \, dz - \int_{t_b/2}^{(t_a+t_b)/2} \sigma_y z \, dz \tag{6.44}$$

$$M_{xy} = -\int_{-(t_a+t_b)/2}^{-t_b/2} \tau_{xy} z \, dz - \int_{-t_b/2}^{t_b/2} \tau_{xy} z \, dz - \int_{t_b/2}^{(t_a+t_b)/2} \tau_{xy} z \, dz$$

which upon substitution of eqns. (6.39) and (6.40) become

$$[M] = -[\bar{Q}]_a \int_{-(t_a+t_b)/2}^{-t_b/2} z[\varepsilon] \, dz - [\bar{Q}]_b \int_{-t_b/2}^{t_b/2} z[\varepsilon] \, dz$$

$$-[\bar{Q}]_a \int_{t_b/2}^{(t_a+t_b)/2} z[\varepsilon] \, dz \tag{6.45}$$

Again $[\varepsilon]$ may be replaced by $([\varepsilon_0] - z[\chi])$ and the formal similarity with eqn. (6.12) noted. On carrying out the integrations the equation

$$[M] = \frac{1}{12}([\bar{Q}]_a(\{t_a + t_b\}^3 - t_b^3) + [\bar{Q}]_b t_b^3)[\chi] \tag{6.46}$$

is obtained, where again the analogy with eqn. (6.12) is apparent.

As an illustration of these general results, consider the simplest of symmetric laminates, namely a three-layered laminate composed of isotropic laminae of equal thickness, t, the middle lamina having different properties from the outer two. An exploded view of such a laminate is depicted in Fig. 6.10. In this case the matrices $[\bar{Q}]_a$ and $[\bar{Q}]_b$ become

$$[\bar{Q}]_a = \begin{bmatrix} \dfrac{E_a}{1-v_a^2} & \dfrac{v_a E_a}{1-v_a^2} & \\ \dfrac{v_a E_a}{1-v_a^2} & \dfrac{E_a}{1-v_a^2} & \\ & & \dfrac{E_a}{2(1+v_a)} \end{bmatrix} \quad (6.47a)$$

$$[\bar{Q}]_b = \begin{bmatrix} \dfrac{E_b}{1-v^2} & \dfrac{v_b E_b}{1-v_b^2} & \\ \dfrac{v_b E_b}{1-v_b^2} & \dfrac{E_b}{1-v_b^2} & \\ & & \dfrac{E_b}{2(1+v_b)} \end{bmatrix} \quad (6.47b)$$

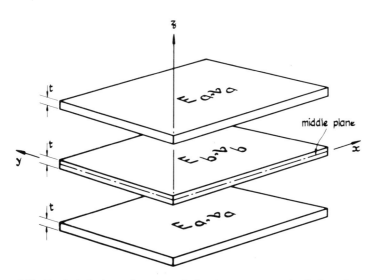

FIG. 6.10. Exploded view of symmetric laminate composed of three isotropic laminae of equal thickness.

where E_a and v_a are Young's modulus and Poisson's ratio for the material of the outer laminae, and E_b and v_b these properties for the middle lamina. On substituting these matrices into eqns. (6.43) and (6.46), and noting that

$$t = \frac{t_a}{2} = t_b$$

it follows that

$$\begin{bmatrix} N_x \\ N_y \\ N_{xy} \end{bmatrix} = t \begin{bmatrix} 2\left(\frac{E_a}{1-v_a^2}\right)+\left(\frac{E_b}{1-v_b^2}\right) & 2\left(\frac{v_a E_a}{1-v_a^2}\right)+\left(\frac{v_b E_b}{1-v_b^2}\right) & \\ 2\left(\frac{v_a E_a}{1-v_a^2}\right)+\left(\frac{v_b E_b}{1-v_b^2}\right) & 2\left(\frac{E_a}{1-v_a^2}\right)+\left(\frac{E_b}{1-v_b^2}\right) & \\ & & \frac{E_a}{(1+v_a)}+\frac{E_b}{2(1+v_b)} \end{bmatrix} \begin{bmatrix} \frac{\partial u_0}{\partial x} \\ \frac{\partial v_0}{\partial y} \\ \frac{\partial u_0}{\partial y}+\frac{\partial v_0}{\partial x} \end{bmatrix}$$

(6.48)

and

$$\begin{bmatrix} M_x \\ M_y \\ M_{xy} \end{bmatrix} = \frac{t^3}{12} \begin{bmatrix} 26\left(\frac{E_a}{1-v_a^2}\right)+\left(\frac{E_b}{1-v_b^2}\right) & 26\left(\frac{v_a E_a}{1-v_a^2}\right)+\left(\frac{v_b E_b}{1-v_b^2}\right) & \\ 26\left(\frac{v_a E_a}{1-v_a^2}\right)+\left(\frac{v_b E_b}{1-v_b^2}\right) & 26\left(\frac{E_a}{1-v_a^2}\right)+\left(\frac{E_b}{1-v_b^2}\right) & \\ & & \frac{13E_a}{(1+v_a)}+\frac{E_b}{2(1+v_b)} \end{bmatrix} \begin{bmatrix} \frac{\partial^2 w}{\partial x^2} \\ \frac{\partial^2 w}{\partial y^2} \\ 2\frac{\partial^2 w}{\partial x \partial y} \end{bmatrix}$$

(6.49)

In this special case it can be noted that there is no interaction between normal and shearing stresses or between bending moments and torques, and similar independence occurs in the slightly more complex symmetric cross-ply laminate in which the isotropic laminae are replaced by similar orthotropic laminae so orientated that their major principal material directions lie alternately along the x and y axes. When the major principal material directions of the orthotropic laminae lie alternately at angles α and $-\alpha$ to either of the axes x or y, the resulting form is termed an angle-ply laminate, a construction which allows normal and shearing interaction to occur.

Finally, if in eqns. (6.48) and (6.49) the material properties of the constituent laminae are identical, it is seen that the equations simply represent an isotropic plate of thickness $3t$.

6.2.2.4 TWO LAYERED NON-SYMMETRIC LAMINATE

Consider finally the stiffness properties of a plate constructed from two generally orthotropic laminae of thicknesses t_a and t_b as shown by the exploded view in Fig. 6.11.

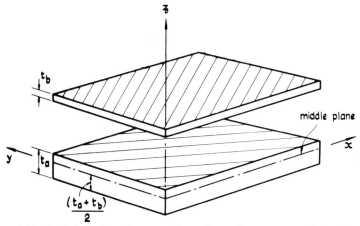

FIG. 6.11. Exploded view of a general two-layered non-symmetric laminate.

If the stress–strain relationships of the bottom and top laminae are given by eqns. (6.39) and (6.40) respectively, then the forces and moments per unit width of the plate may be obtained by combining the effects of the two laminae in a manner analogous to that presented in eqns. (6.13) and (6.14) for a non-symmetrically laminated beam. The axial and shearing forces per unit width of the laminate will thus be given by

$$N_x = \int_{-(t_a+t_b)/2}^{(t_a-t_b)/2} \sigma_x \, dz + \int_{(t_a-t_b)/2}^{(t_a+t_b)/2} \sigma_x \, dz$$

$$N_y = \int_{-(t_a+t_b)/2}^{(t_a-t_b)/2} \sigma_y \, dz + \int_{(t_a-t_b)/2}^{(t_a+t_b)/2} \sigma_y \, dz$$

$$N_{xy} = \int_{-(t_a+t_b)/2}^{(t_a-t_b)/2} \tau_{xy} \, dz + \int_{(t_a-t_b)/2}^{(t_a+t_b)/2} \tau_{xy} \, dz \qquad (6.50)$$

where the first term in each equation describes the bottom layer of thickness t_a and the second term the top layer of thickness t_b, which upon substitution of eqns. (6.39) and (6.40) may be written

$$[N] = [\bar{Q}]_a \int_{-(t_a+t_b)/2}^{(t_a-t_b)/2} [\varepsilon]\,dz + [\bar{Q}]_b \int_{(t_a-t_b)/2}^{(t_a+t_b)/2} [\varepsilon]\,dz \quad (6.51)$$

where, as before,

$$[\varepsilon] = [\varepsilon_0] - z[\chi]$$

At this point it is again pertinent to draw attention to the similarity in form of eqns. (6.51) and (6.13), which quantifies the axial force in a two-layered beam.

Now, on performing the integrations in eqns. (6.51), the in-plane forces in the laminate are given by

$$[N] = ([\bar{Q}]_a t_a + [\bar{Q}]_b t_b)[\varepsilon_0] + ([\bar{Q}]_a - [\bar{Q}]_b)\frac{t_a t_b}{2}[\chi] \quad (6.52)$$

where formal similarity with the final form of eqn. (6.13) is strikingly apparent.

In the same manner the equations for the bending moments and torques per unit width of the laminate may be expressed as

$$M_x = -\int_{-(t_a+t_b)/2}^{(t_a-t_b)/2} \sigma_x z\,dz - \int_{(t_a-t_b)/2}^{(t_a+t_b)/2} \sigma_x z\,dz$$

$$M_y = -\int_{-(t_a+t_b)/2}^{(t_a-t_b)/2} \sigma_y z\,dz - \int_{(t_a-t_b)/2}^{(t_a+t_b)/2} \sigma_y z\,dz \quad (6.53)$$

$$M_{xy} = -\int_{-(t_a+t_b)/2}^{(t_a-t_b)/2} \tau_{xy} z\,dz - \int_{(t_a-t_b)/2}^{(t_a+t_b)/2} \tau_{xy} z\,dz$$

or in matrix form,

$$[M] = -[\bar{Q}]_a \int_{-(t_a+t_b)/2}^{(t_a-t_b)/2} z[\varepsilon]\,dz - [\bar{Q}]_b \int_{(t_a-t_b)/2}^{(t_a+t_b)/2} z[\varepsilon]\,dz \quad (6.54)$$

Upon the performing of the integrations eqns. (6.54) become

$$[M] = ([\bar{Q}]_a - [\bar{Q}]_b) \frac{t_a t_b}{2} [\varepsilon_0]$$

$$+ \frac{1}{12}([\bar{Q}]_a t_a [t_a^2 + 3t_b^2] + [\bar{Q}]_b t_b [3t_a^2 + t_b^2])[\chi] \quad (6.55)$$

where again similarity with the behaviour of the non-symmetrically laminated beam, as described in eqn. (6.14), is noticed.

The coupling that exists between the in-plane and lateral deformations in non-symmetic composites is again shown in eqns. (6.52) and 6.55).

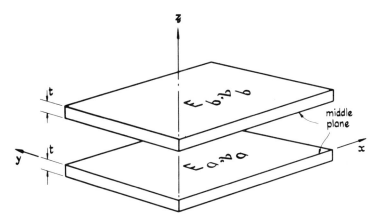

FIG. 6.12. Exploded view of non-symmetric laminate composed of two isotropic laminae of equal thickness.

Let these results now be applied to the determination of the stiffness properties of a non-symmetric laminate composed of two differing isotropic laminae of equal thickness; t, such a laminate being shown in Fig. 6.12. Here the matrices $[\bar{Q}]_a$ and $[\bar{Q}]_b$ are as given by eqns. (6.47); noting that

$$t = t_a = t_b$$

and substituting eqns. (6.47) into eqns. (6.52) and (6.55) the following equations are obtained for the axial and flexural relationships:

$$\begin{bmatrix} N_x \\ N_y \\ N_{xy} \\ M_x \\ M_y \\ M_{xy} \end{bmatrix} = \begin{bmatrix} A & B \\ & \\ B & D \end{bmatrix} \begin{bmatrix} \dfrac{\partial u_0}{\partial x} \\ \dfrac{\partial v_0}{\partial y} \\ \dfrac{\partial u_0}{\partial y} + \dfrac{\partial v_0}{\partial x} \\ \dfrac{\partial^2 w}{\partial x^2} \\ \dfrac{\partial^2 w}{\partial y^2} \\ 2\dfrac{\partial^2 w}{\partial x \partial y} \end{bmatrix} \quad (6.56)$$

where

$$[A] = t \begin{bmatrix} \dfrac{E_a}{1-v_a^2} + \dfrac{E_b}{1-v_b^2} & \dfrac{v_a E_a}{1-v_a^2} + \dfrac{v_b E_b}{1-v_b^2} & \\ \dfrac{v_a E_a}{1-v_a^2} + \dfrac{v_b E_b}{1-v_b^2} & \dfrac{E_a}{1-v_a^2} + \dfrac{E_b}{1-v_b^2} & \\ & & \dfrac{E_a}{2(1+v_a)} + \dfrac{E_b}{2(1+v_b)} \end{bmatrix} \quad (6.57a)$$

$$[D] = \dfrac{t^2}{3}[A] \quad (6.57b)$$

and

$$[B] = \dfrac{t^2}{2} \begin{bmatrix} \dfrac{E_a}{1-v_a^2} - \dfrac{E_b}{1-v_b^2} & \dfrac{v_a E_a}{1-v_a^2} - \dfrac{v_b E_b}{1-v_b^2} & \\ \dfrac{v_a E_a}{1-v_a^2} - \dfrac{v_b E_b}{1-v_b^2} & \dfrac{E_a}{1-v_a^2} - \dfrac{E_b}{1-v_b^2} & \\ & & \dfrac{E_a}{2(1+v_a)} - \dfrac{E_b}{2(1+v_b)} \end{bmatrix} \quad (6.57c)$$

Similar results may be obtained for such a laminate composed of two orthotropic laminae, while again, if in eqn. (6.56) the two laminae are identical, it is observed that the equations reduce to those describing the behaviour of an isotropic plate of thickness $2t$, the coupling matrix $[B]$ in this case vanishing.

6.2.2.5 General Equations for Laminated Plates

The results presented for the elementary symmetric and non-symmetric laminates show that the general relationships between the planar and flexural stress resultants and the corresponding strains may be expressed as

$$\begin{bmatrix} N \\ M \end{bmatrix} = \begin{bmatrix} A & B \\ B & D \end{bmatrix} \begin{bmatrix} \varepsilon_0 \\ \chi \end{bmatrix} \tag{6.58}$$

These equations are the plate equivalents of eqn. (6.16) for a laminated beam, which indeed may be viewed as the special case of eqn. (6.58) when the matrices A, B and D contain one element only.

The examples of laminated composites discussed so far contain only a minimum number of laminae in order to outline the fundamental nature of laminate behaviour. In practice a laminate may consist of a large number of separate laminae and therefore it is necessary to generalise the equations so far developed for simple laminates to enable the stiffness properties of multilayered composites to be determined.[1,3,9,10]

Consider such a multilayered laminate, a section through which is depicted in Fig. 6.13. On examining eqns. (6.41) and (6.44) for the three-

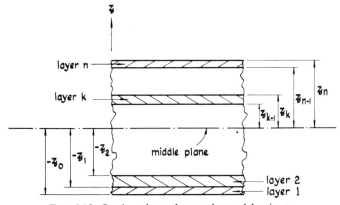

Fig. 6.13. Section through an n-layered laminate.

layered symmetric laminate, it can be seen that each equation contains three terms, while in the similar eqns. (6.50) and (6.53) for the two-layered non-symmetric laminate each equation contains two terms. Thus it can be deduced that for an n-layered laminate the equations will simply be extended so that each contains n terms, i.e. as many terms as there are laminae.

For instance,

$$N_x = \int_{z_0}^{z_1} \sigma_x \, dz + \int_{z_1}^{z_2} \sigma_x \, dz + \ldots + \int_{z_{k-1}}^{z_k} \sigma_x \, dz + \ldots + \int_{z_{n-1}}^{z_n} \sigma_x \, dz$$

$$N_y = \int_{z_0}^{z_1} \sigma_y \, dz + \int_{z_1}^{z_2} \sigma_y \, dz + \ldots + \int_{z_{k-1}}^{z_k} \sigma_y \, dz + \ldots + \int_{z_{n-1}}^{z_n} \sigma_y \, dz$$

$$N_{xy} = \int_{z_0}^{z_1} \tau_{xy} \, dz + \int_{z_1}^{z_2} \tau_{xy} \, dz + \ldots + \int_{z_{k-1}}^{z_k} \tau_{xy} \, dz + \ldots + \int_{z_{n-1}}^{z_n} \tau_{xy} \, dz$$

(6.59)

and

$$M_x = -\int_{z_0}^{z_1} \sigma_x z \, dz - \int_{z_1}^{z_2} \sigma_x z \, dz - \ldots - \int_{z_{k-1}}^{z_k} \sigma_x z \, dz - \ldots - \int_{z_{n-1}}^{z_n} \sigma_x z \, dz$$

$$M_y = -\int_{z_0}^{z_1} \sigma_y z \, dz - \int_{z_1}^{z_2} \sigma_y z \, dz - \ldots - \int_{z_{k-1}}^{z_k} \sigma_y z \, dz - \ldots - \int_{z_{n-1}}^{z_n} \sigma_y z \, dz$$

$$M_{xy} = -\int_{z_0}^{z_1} \tau_{xy} z \, dz - \int_{z_1}^{z_2} \tau_{xy} z \, dz - \ldots - \int_{z_{k-1}}^{z_k} \tau_{xy} z \, dz - \ldots - \int_{z_{n-1}}^{z_n} \tau_{xy} z \, dz$$

(6.60)

Now the relationships between stress and strain for any particular lamina k may be written as

$$\begin{bmatrix} \sigma_x \\ \sigma_y \\ \tau_{xy} \end{bmatrix} = [\bar{Q}]_k \begin{bmatrix} \varepsilon_x \\ \varepsilon_y \\ \gamma_{xy} \end{bmatrix} \quad (6.61)$$

being of the same form as eqns. (6.39) and (6.40). Thus on substituting relationships (6.61) into eqns. (6.59) and (6.60), the stress resultants can be written

$$[N] = [\bar{Q}]_1 \int_{z_0}^{z_1} [\varepsilon]\,dz + [\bar{Q}]_2 \int_{z_1}^{z_2} [\varepsilon]\,dz + \ldots$$
$$+ [\bar{Q}]_k \int_{z_{k-1}}^{z_k} [\varepsilon]\,dz + \ldots + [\bar{Q}]_n \int_{z_{n-1}}^{z_n} [\varepsilon]\,dz \tag{6.62}$$

and

$$[M] = -[\bar{Q}]_1 \int_{z_0}^{z_1} z[\varepsilon]\,dz - [\bar{Q}]_2 \int_{z_1}^{z_2} z[\varepsilon]\,dz - \ldots$$
$$- [\bar{Q}]_k \int_{z_{k-1}}^{z_k} z[\varepsilon]\,dz - \ldots - [\bar{Q}]_n \int_{z_{n-1}}^{z_n} z[\varepsilon]\,dz \tag{6.63}$$

Now, by writing $[\varepsilon]$ as $([\varepsilon_0] - z[\chi])$, and carrying out the integrations, eqns. (6.62) and (6.63) become

$$[N] = \left\{ \sum_{k=1}^{n} (z_k - z_{k-1})[\bar{Q}]_k \right\} [\varepsilon_0]$$
$$- \left\{ \frac{1}{2} \sum_{k=1}^{n} (z_k^2 - z_{k-1}^2)[\bar{Q}]_k \right\} [\chi] \tag{6.64}$$

and

$$[M] = -\left\{ \frac{1}{2} \sum_{k=1}^{n} (z_k^2 - z_{k-1}^2)[\bar{Q}]_k \right\} [\varepsilon_0]$$
$$+ \left\{ \frac{1}{3} \sum_{k=1}^{n} (z_k^3 - z_{k-1}^3)[\bar{Q}]_k \right\} [\chi] \tag{6.65}$$

Thus the generalised stress resultant to strain relationships for an n-layered laminate may be written as

$$\begin{bmatrix} N_x \\ N_y \\ N_{xy} \end{bmatrix} = \begin{bmatrix} A_{11} & A_{12} & A_{13} \\ A_{12} & A_{22} & A_{23} \\ A_{13} & A_{23} & A_{33} \end{bmatrix} \begin{bmatrix} \dfrac{\partial u_0}{\partial x} \\ \dfrac{\partial v_0}{\partial y} \\ \dfrac{\partial u_0}{\partial y} + \dfrac{\partial v_0}{\partial x} \end{bmatrix} + \begin{bmatrix} B_{11} & B_{12} & B_{13} \\ B_{12} & B_{22} & B_{23} \\ B_{13} & B_{23} & B_{33} \end{bmatrix} \begin{bmatrix} \dfrac{\partial^2 w}{\partial x^2} \\ \dfrac{\partial^2 w}{\partial y^2} \\ 2\dfrac{\partial^2 w}{\partial x \partial y} \end{bmatrix}$$

$$\begin{bmatrix} M_x \\ M_y \\ M_{xy} \end{bmatrix} = \begin{bmatrix} B_{11} & B_{12} & B_{13} \\ B_{12} & B_{22} & B_{23} \\ B_{13} & B_{23} & B_{33} \end{bmatrix} \begin{bmatrix} \dfrac{\partial u_0}{\partial x} \\ \dfrac{\partial v_0}{\partial y} \\ \dfrac{\partial u_0}{\partial y} + \dfrac{\partial v_0}{\partial x} \end{bmatrix} + \begin{bmatrix} D_{11} & D_{12} & D_{13} \\ D_{12} & D_{22} & D_{23} \\ D_{13} & D_{23} & D_{33} \end{bmatrix} \begin{bmatrix} \dfrac{\partial^2 w}{\partial x^2} \\ \dfrac{\partial^2 w}{\partial y^2} \\ 2\dfrac{\partial^2 w}{\partial x \partial y} \end{bmatrix}$$

(6.66)

where

$$A_{ij} = \sum_{k=1}^{n} (z_k - z_{k-1})[\bar{Q}]_k$$

$$B_{ij} = -\frac{1}{2} \sum_{k=1}^{n} (z_k^2 - z_{k-1}^2)[\bar{Q}]_k$$

$$D_{ij} = \frac{1}{3} \sum_{k=1}^{n} (z_k^3 - z_{k-1}^3)[\bar{Q}]_k$$

all three matrices $[A]$, $[B]$ and $[D]$ being seen to possess symmetry.

The corresponding results for an n-layered beam are precisely the same, except that the matrices $[A]$, $[B]$ and $[D]$ contain only one element.

In the analysis of laminated composites, the inclusion of normal stress–shearing stress coupling and bending–membrane interaction is essential, as failure to consider these effects in an analysis can lead to non-conservative results.[3,10]

6.3 STRENGTH CHARACTERISTICS OF LAMINATED COMPOSITES

As with laminate stiffness characteristics, the basic parameters required to develop a description of laminate strengths are the strength characteristics of a single lamina as outlined in the previous chapter.

The strength of the composite laminate will, like the stiffness characteristics, depend on a number of variables such as the quantity, orientation and thickness of the constituent laminae, as well as on the voids content and the temperature at which the lamination is carried out. The geometrical variables affect the strength analogously to the way in which they affect the stiffness properties, while the thermal effect tends to produce initial strains in each lamina which, if the laminating temperature is considerably higher than the temperature of the environment in which the laminate is to be used, should be added to the subsequent strains produced by the application of stress.[1,3,11,12]

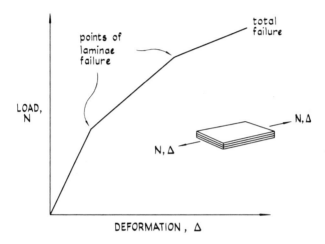

FIG. 6.14. Load–deformation characteristics of a laminate to failure.

The fundamental difference between the failure of a laminated composite and an individual lamina is that, whereas in a lamina the load–deformation relationship to failure is reasonably linear and continuous, that of the laminate is generally of a discontinuous nature, due to the fact that total failure occurs by progressive failure of the individual laminae. This form of failure occurs since, because of the different orientation of the laminae, in any particular direction in which the strength is measured each lamina will generally have a strength of differing magnitude. Thus when the weakest lamina fails, the load is carried by those remaining with a consequent loss of stiffness. Further increase of load initiates failure of the next weakest lamina, and this process is continued until total failure occurs.[1,3,13] Thus the load–deformation characteristics of a laminate to failure will exhibit the form shown in Fig. 6.14.

6.3.1 Strength Analysis and Failure Criteria

To demonstrate the manner in which laminate strengths may be calculated, consider a composite whose laminations are symmetrical about the middle plane (so that no coupling exists between in-plane and flexural action), subjected to uniaxial tension in the x-direction, as shown in Fig. 6.15.

The load–deformation behaviour of such a material may be expressed by the equations

$$\begin{bmatrix} N_x \\ 0 \\ 0 \end{bmatrix} = \begin{bmatrix} A_{11} & A_{12} & A_{13} \\ A_{12} & A_{22} & A_{23} \\ A_{13} & A_{23} & A_{33} \end{bmatrix} \begin{bmatrix} \dfrac{\partial u_0}{\partial x} \\ \dfrac{\partial v_0}{\partial y} \\ \dfrac{\partial u_0}{\partial y} + \dfrac{\partial v_0}{\partial x} \end{bmatrix} = [A][\varepsilon_0] \quad (6.67)$$

where in general A_{13} and A_{23} will be non-zero due to coupling between normal and shearing effects that can occur in the plane of each lamina.

By inverting $[A]$ these equations may be rewritten so as to express the strains in terms of the uniaxial tension, N_x, thus:

$$\begin{bmatrix} \dfrac{\partial u_0}{\partial x} \\ \dfrac{\partial v_0}{\partial y} \\ \dfrac{\partial u_0}{\partial y} + \dfrac{\partial v_0}{\partial x} \end{bmatrix} = \begin{bmatrix} A'_{11} & A'_{12} & A'_{13} \\ A'_{12} & A'_{22} & A'_{23} \\ A'_{13} & A'_{23} & A'_{33} \end{bmatrix} \begin{bmatrix} N_x \\ 0 \\ 0 \end{bmatrix} = [A]^{-1} \begin{bmatrix} N_x \\ 0 \\ 0 \end{bmatrix} \quad (6.68)$$

or more simply

$$\frac{\partial u_0}{\partial x} = A'_{11} N_x$$

$$\frac{\partial v_0}{\partial y} = A'_{12} N_x$$

$$\frac{\partial u_0}{\partial y} + \frac{\partial v_0}{\partial x} = A'_{13} N_x$$

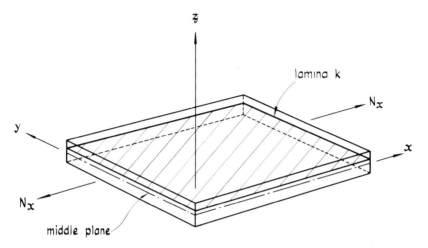

FIG. 6.15. Laminate under uniaxial tension, N_x.

These strains are the same for every lamina in the composite and hence the stresses in any particular lamina k may be expressed as

$$\begin{bmatrix} \sigma_x \\ \sigma_y \\ \tau_{xy} \end{bmatrix} = \begin{bmatrix} \bar{Q}_{11} \bar{Q}_{12} \bar{Q}_{13} \\ \bar{Q}_{12} \bar{Q}_{22} \bar{Q}_{23} \\ \bar{Q}_{13} \bar{Q}_{23} \bar{Q}_{33} \end{bmatrix}_k \begin{bmatrix} A'_{11} N_x \\ A'_{12} N_x \\ A'_{13} N_x \end{bmatrix} \quad (6.69)$$

where the coefficients $\bar{Q}_{ij}$ are the stiffness coefficients of lamina k related to the x- and y-directions.

The stresses and strains in the principal material directions 1 and 2 of the lamina, if these be different from the x- and y-directions, may now be obtained from the standard transformations, after which process they may be substituted into one of the failure criteria discussed in the previous chapter to determine the value of N_x required to cause failure of the weakest lamina. The process can then be repeated until total failure of the laminate takes place.

Strengths of laminates of different forms and/or under different loading systems may be considered in a similar manner.

As discussed in Chapter 4, of the failure criteria considered, experimental evidence suggests that the Tsai–Hill failure criterion predicts most accurately the actual strength of glass fibre reinforced laminates.[14]

6.4 THE EFFECT OF INTERLAMINAR STRESSES

In all the investigations up till now the only stresses that have been considered operative in the discussion of laminate behaviour are the three in-plane stresses, σ_x, σ_y, τ_{xy} acting on each lamina (Fig. 6.16a). Laminates have been simply considered as a stack of such laminae, the stress resultants acting on the laminate being obtained from the cumulative effect of the behaviour of the individual laminae.

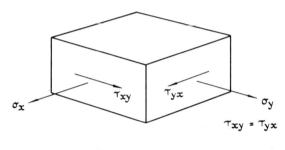

a) In-plane stresses

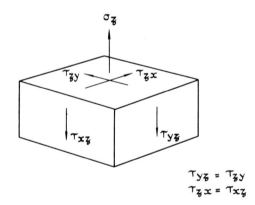

b) Interlaminar stresses

FIG. 6.16. In-plane and interlaminar stresses induced in a laminate.

There are, however, three further stresses which must be considered as their action is to tend to cause debonding of the individual laminae. These additional stresses, the normal stress σ_z and the two shearing stresses τ_{yz} and τ_{zx}, are shown in Fig. 6.16b.

The consideration of vertical equilibrium in a beam alone is sufficient to demonstrate the presence of vertical shearing stresses with their corresponding complementary horizontal stresses, and it is the effect of this horizontal action that can, if too severe, cause sliding of one lamina relative to another. Since the glass fibres are all in horizontal planes, there are interlaminar regions consisting only of the resin matrix and thus the allowable interlaminar shearing stress will be primarily dependent upon the shearing strength of the matrix if delamination is to be avoided.

It should also be noted that in the small region very close to a free edge of a laminated plate, the lamination theory presented in this chapter does not apply. In these areas the values of the interlaminar shearing stresses τ_{yz}, τ_{zx} and also of the transverse normal stress σ_z may be very high, their magnitudes being dependent in part upon the orientations and stacking sequence of the individual laminae. The calculation of these stresses in such regions is not straightforward,[15-17] and numerical methods such as finite differences or finite elements are usually invoked if their determination is considered essential.[18-20] Because of this phenomenon, special consideration should be given to the design of laminates in the neighbourhood of free edges or holes.

REFERENCES

1. AGARWAL, B. D. and BROUTMAN, L. J., *Analysis and Performance of Fiber Composites*, 1980, John Wiley, New York.
2. BROUTMAN, L. J. and KROCK, R. H., *Modern Composite Materials*, 1967, Addison-Wesley, Reading, Mass.
3. JONES, R. M., *Mechanics of Composite Materials*, 1975, McGraw-Hill, New York.
4. PIGGOTT, M. R., *Load Bearing Fibre Composites*, 1980, Pergamon Press Ltd, Oxford.
5. SALAMON, N. J., An assessment of the interlaminar stress problem in laminated composites, *J. Composite Materials Supplement*, **14**, 1980, 177.
6. BENHAM, P. P. and WARNOCK, F. V., *Mechanics of Solids and Structures*, 1980, Pitman, London.
7. TIMOSHENKO, S. P., *Strength of Materials, Part 1*, 3rd edn., 1980, Van Nostrand Reinhold, Princeton, NJ.
8. TIMOSHENKO, S. P. and WOINOWSKY-KRIEGER, S., *Theory of Plates and Shells*, 2nd edn., 1959, McGraw-Hill, New York.
9. VINSON, J. R. and CHOU, T. W., *Composite Materials and their Use in Structures*, 1975, Applied Science Publishers, London.

10. ASHTON, J. E., HALPIN, J. C. and PETIT, P. H., *Primer on Composite Materials: Analysis*, 1969, Technomic Publishing, Stanford, Conn.
11. HAHN, H. T. Residual stresses in polymer matrix composite laminates, *J. Composite Materials*, **10**, Oct. 1976, 266.
12. HAHN, H. T. and PAGANO, N. J., Curing stresses in composite laminates, *J. Composite Materials*, **9**, Jan. 1975, 91.
13. HAHN, H. T. and TSAI, S. W., On the behaviour of composite laminates after initial failure, *J. Composite Materials*, **8**, July 1974, 288.
14. OWEN, M. J. and RICE, D. J., Biaxial strength behaviour of glass fabric reinforced polyester resins, *Composites*, **12**, Jan. 1981, 13.
15. PAGANO, N. J., On the calculation of interlaminar normal stress in composite laminate, *J. Composite Materials*, **8**, Jan. 1974, 65.
16. PIPES, R. B. and PAGANO, N. J., Interlaminar stresses in composite laminates — an approximate elasticity solution, *J. Applied Mechanics*, **41**, 1974, 668.
17. PIPES, R. B., Boundary layer effects in composite materials, *Fibre Science and Technology*, **13**, 1980, 49.
18. ALTUS, E., ROTEM, A. and SHMUELI, M., Free edge effect in angle ply laminates — a new three dimensional finite difference solution, *J. Composite Materials*, **14**, 1980, 21.
19. PIPES, R. B. and PAGANO, N. J., Interlaminar stresses in composite laminates under uniform axial extension, *J. Composite Materials*, **4**, 1970, 538.
20. SPILKER, R. L. and CHOU, S. C., Edge effects in symmetric composite laminates: importance of satisfying the traction-free-edge condition, *J. Composite Materials*, **14**, 1980, 2.

CHAPTER 7

The Theoretical and Measured Properties of Glass-reinforced Composites

7.1 INTRODUCTION

Since the analytical results for composite materials developed in the last three chapters have been of necessity built up from a number of simplifying assumptions, it is essential that their accuracy and value be assessed in the light of direct experimental evidence.

Such evidence should suggest whether a particular theoretical treatment can be used confidently in a design situation, or whether a design should be governed by experimental data. It is the object of this chapter to examine the theoretical and experimentally determined properties of a range of glass fibre-reinforced composites for the various stiffness and strength characteristics described previously.

Prime attention will be devoted to the consideration of continuously reinforced unidirectional and multidirectional laminates of various fibre contents, after which laminates constructed from discontinuous fibres will be briefly discussed.

It should be noted that the experimental results quoted are obtained from short term tests at ambient temperatures, or in other words any time- or temperature-dependent behaviour of the materials is not considered.

7.2 CONTINUOUSLY REINFORCED LAMINATES

Glass fibre-reinforced laminates constructed using continuous reinforcement may be divided into two classes, namely

(a) unidirectionally reinforced composites, and
(b) multidirectionally reinforced composites.

Unidirectionally reinforced laminates are generally the analytically simpler of the two, and furthermore if the individual laminae of the composite are identical in composition and orientation then many of the properties of the laminate may be theoretically obtained by treating it as an equivalent single lamina.

Both the unidirectionally and the multidirectionally reinforced laminates that will be investigated contain, as their basic constituents, standard 'E' glass unidirectional rovings and standard commercially produced polyester resin that cures at room temperature.[1,2] The glass fibres are assumed to possess a constant value of modulus of elasticity of 72·4 kN/mm^2 to failure, the tensile failure strength in the perfect condition being considered to be 3·45 kN/mm^2, and a constant value of Poisson's ratio of 0·2. While the glass fibres may be considered to possess reasonably standard property values, the characteristics of polyester resins exhibit far more variation and hence the values of the resin properties used in the investigations are those obtained from tests on the material in both tension and compression.[1,2] The stiffness and strength properties of the glass fibre reinforcement and the polyester resin thus obtained are presented in Fig. 7.1.

The experimental results further presented are those furnished from tests on reinforced laminates manufactured using the standard hand lay-up technique.[1,2] The voids ratio of the resulting composites is about 3%, a relatively high value in comparison with that which can be attained in closed mould manufacturing processes.

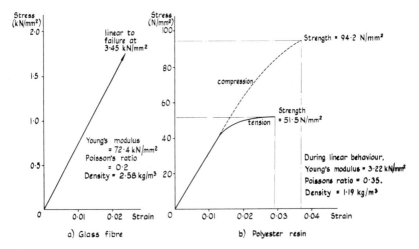

FIG. 7.1. Measured properties of glass fibre and polyester resin.

The theoretical and measured properties of glass-reinforced composites 173

Two basic forms of test are considered for the determination of the stiffness and strength properties of the various laminates chosen for investigation, these being

(a) tensile and compressive tests, and
(b) flexural tests.

The tensile and compressive tests in the directions of the principal material axes are used to measure the values of Young's moduli and Poisson's ratios relative to these directions and also the corresponding tensile and compressive strengths. The modulus of rigidity relative to the principal material axes can be obtained from similar tests in which the stress is applied along any direction which is not a principal axis of the material, as depicted in Fig. 7.2. That this is so can be deduced from the first of eqns. (4.96) which, if only σ_x were applied, would yield

$$\varepsilon_x = \bar{S}_{11} \sigma_x$$
$$= \{S_{11} \cos^4\theta + S_{22} \sin^4\theta + (2S_{12} + S_{33}) \sin^2\theta \cos^2\theta\} \sigma_x \quad (7.1)$$

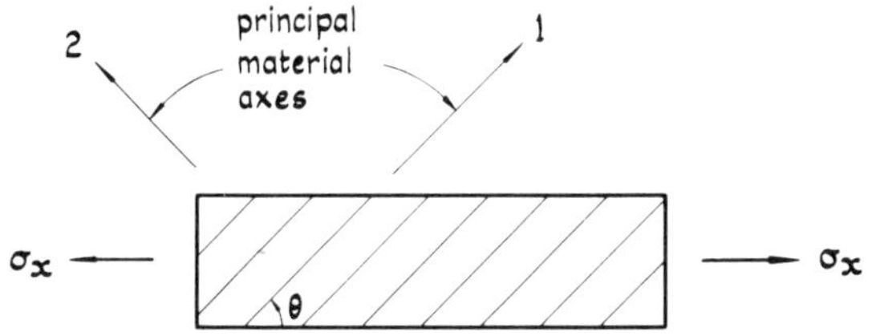

FIG. 7.2. Test used to determine G_{12}.

Since ε_x can be measured and S_{11}, S_{22} and S_{12} are functions of the Young's moduli and Poisson's ratios relative to the principal material directions, the value of S_{33}, which is the reciprocal of the required modulus of rigidity, can be calculated. Denoting $\bar{S}_{11}$ by $1/E_x$, in the special case where $\theta = 45°$, eqn. (7.1) reduces to

$$\frac{1}{E_x} = \frac{1}{4}\left(\frac{1}{E_1} + \frac{1}{E_2} - \frac{2v_{12}}{E_1} + \frac{1}{G_{12}}\right) \quad (7.2a)$$

from which

$$\frac{1}{G_{12}} = \frac{4}{E_x} - \frac{(1-2v_{12})}{E_1} - \frac{1}{E_2} \quad (7.2b)$$

The flexural tests can be used to examine the flexural properties and the interlaminar shearing strengths of the laminates.

7.2.1 Unidirectionally Continuously Reinforced Laminates

7.2.1.1 PROPERTIES OBTAINED FROM TENSILE AND COMPRESSIVE TESTS
Carefully executed tensile and compressive tests can produce a wealth of data regarding the behaviour and properties of a material. In the case of unidirectionally continuously reinforced laminates such tests may be carried out in both the longitudinal and transverse directions to obtain the respective stiffness and strength properties, while tests performed on specimens with the fibre direction inclined to the axis of loading can illuminate the shearing characteristics of the material.

A spectrum of results, both analytical and experimental, will be presented for these various stiffness and strength values in laminates composed of a number of identical laminae and containing fibre volume fractions ranging from 0·27 to 0·64, each experimental result being the mean of five tests.[1,2]

In the prediction of the stiffness and strength characteristics of such laminates, recourse will be made to the micromechanical analyses of a single lamina described in Chapter 5 and to the constituent properties stated in Fig. 7.1.

Consider firstly the stiffness properties E_1, E_2, v_{12}, v_{21} and G_{12} for a laminate constructed from a number of identical unidirectional laminae and with a fibre volume fraction, V_f, of 0·27.

Using the initial constant slope of the stress–strain curve for the polyester resin, the value of Young's modulus in the direction of the fibres, E_1, may be expressed by eqn. (5.6) as

$$E_1 = E_f V_f + E_m V_m = E_f V_f + E_m (1 - V_f)$$
$$= (72·4 \times 0·27) + (3·22 \times 0·73)$$
$$= 19·55 + 2·35$$
$$= 21·90 \text{ kN/mm}^2$$

where 72·4 kN/mm² and 3·22 kN/mm² are the experimentally measured values of Young's modulus for the glass fibres and resin respectively.

Inspection of the contributions of the component parts to this result shows its high dependence on the stiffness of the fibres.

The value of Poisson's ratio v_{12} may be similarly obtained from eqn. (5.11) as

$$v_{12} = v_f V_f + v_m V_m = v_f V_f + v_m (1 - V_f)$$
$$= (0·2 \times 0·27) + (0·35 \times 0·73)$$
$$= 0·31$$

The theoretical and measured properties of glass-reinforced composites 175

the values 0·2 and 0·35 being the experimentally determined Poisson's ratio values for the glass fibres and resin respectively.

Use of eqn. (5.17a) or (5.17b) enables the value of Young's modulus in the direction transverse to the fibre axis, E_2, to be determined, thus:

$$E_2 = \frac{E_f E_m}{E_m V_f + E_f V_m} = \frac{E_f E_m}{E_m V_f + E_f (1 - V_f)}$$

$$= \frac{72 \cdot 4 \times 3 \cdot 22}{(3 \cdot 22 \times 0 \cdot 27) + (72 \cdot 4 \times 0 \cdot 73)}$$

$$= 4 \cdot 34 \text{ kN/mm}^2$$

The relationship given by eqn. (4.28), i.e.

$$\frac{v_{12}}{E_1} = \frac{v_{21}}{E_2}$$

may now be invoked if required to predict the value of v_{21}. Thus

$$v_{21} = v_{12} \frac{E_2}{E_1}$$

$$= 0 \cdot 31 \times \frac{4 \cdot 34}{21 \cdot 90}$$

$$= 0 \cdot 061$$

Lastly, the value of the modulus of rigidity, G_{12}, may be determined using eqn. (5.24a) or (5.24b):

$$G_{12} = \frac{G_f G_m}{G_m V_f + G_f V_m} = \frac{G_f G_m}{G_m V_f + G_f (1 - V_f)}$$

Since

$$G_f = \frac{E_f}{2(1 + v_f)}$$

$$= \frac{72 \cdot 4}{2(1 + 0 \cdot 2)} = 30 \cdot 17 \text{ kN/mm}^2$$

and

$$G_m = \frac{E_m}{2(1+v_m)}$$

$$= \frac{3\cdot 22}{2(1+0\cdot 35)} = 1\cdot 19 \text{ kN/mm}^2$$

$$G_{12} = \frac{30\cdot 17 \times 1\cdot 19}{(1\cdot 19 \times 0\cdot 27)+(30\cdot 17 \times 0\cdot 73)}$$

$$= 1\cdot 61 \text{ kN/mm}^2$$

When attempting to determine an analytical result for the tensile strength in the fibre direction, it is immediately realised that such a prediction will depend largely on the value assigned to the ultimate strength of the glass fibre reinforcement. Although in the newly drawn state 'E' glass fibre can have a tensile strength of the order of $3\cdot 45\text{ kN/mm}^2$, as depicted in Fig. 7.1, after formation into rovings or fabrics, and after manipulation in the composite manufacturing processes, this strength is considerably reduced. The magnitude of the reduction, which will vary from fibre to fibre, will be dependent upon the voids content, the nature and degree of handling, and also on the amount of fibre misalignment produced during fabrication, and such a strength reduction can only be ascertained from the results of direct experimentation.

Tests on unidirectional continuous glass fibre-reinforced polyesters produced by hand lay-up procedures tend to show that when stressed in the direction of the fibres, tensile failure of such composites occurs at a strain of about 2%, i.e. at a value of about $0\cdot 02E_1$, as long as the failure is fibre controlled, a condition that invariably occurs in a practical situation.

It is possible that the reinforcement strengths in, for example, filament wound composites may be slightly greater due to the smaller degree of handling incurred in the manufacturing process.

Assuming that failure occurs at a strain of 2%, the ultimate strength of the reinforcement is

$$\sigma_{fu} = 0\cdot 02 \times 72\cdot 4$$
$$= 1\cdot 448 \text{ kN/mm}^2$$

a result of the same order as that observed by Ainsworth.[3] Hence from eqn. (5.34) the ultimate tensile strength, X_t, in the direction of the fibres

is given by
$$X_t = \sigma_{fu}V_f + \sigma_{mf}V_m = \sigma_{fu}V_f + \sigma_{mf}(1-V_f)$$
where σ_{mf} is the stress in the matrix at failure of the fibres, i.e. at a strain value of 0·02.

Considering again the laminate composed of identical unidirectional laminae and with $V_f = 0·27$,

$$\begin{aligned}X_t &= (1·448 \times 0·27) + ((0·02 \times 3·22) \times 0·73) \\ &= 0·391 + 0·047 \\ &= 0·438 \text{ kN/mm}^2 \\ &= 438 \text{ N/mm}^2\end{aligned}$$

assuming Young's modulus of the resin to remain constant up to failure of the composite.

The ultimate tensile strength in the direction transverse to the fibres, Y_t, is highly dependent upon the ultimate tensile strength of the resin matrix, although due to voids and stress concentrations that can occur at the fibre–matrix interfaces it is usually less than this value. The reduction in strength is variable, although a loss of the order of 50% is not uncommon. On this basis the value of Y_t would be given by

$$\begin{aligned}Y_t &= \frac{1}{2} \times 51·5 \\ &= 25·8 \text{ N/mm}^2\end{aligned}$$

where 51·5 N/mm^2 is the experimentally determined tensile strength of the resin.

A theoretical estimation of the compressive strength of the composite in the fibre direction can be attempted by determining the smaller of the two values given by an 'extensional' and a 'shear' mode analysis (eqn. (5.38a) and (5.39a)).

Thus, again examining the unidirectionally reinforced laminate with $V_f = 0·27$, for the 'extensional' mode

$$\begin{aligned}X_c &= 2V_f \left[\frac{V_f E_m E_f}{3(1-V_f)} \right]^{1/2} \\ &= 2 \times 0·27 \times \left[\frac{0·27 \times 3·22 \times 72·4}{3 \times (1-0·27)} \right]^{1/2} \\ &= 2·895 \text{ kN/mm}^2 \\ &= 2895 \text{ N/mm}^2\end{aligned}$$

while for the 'shear' mode

$$X_c = \frac{G_m}{1-V_f}$$

$$= \frac{E_m}{2(1+v_m)(1-V_f)}$$

$$= \frac{3 \cdot 22}{2(1+0 \cdot 35)(1-0 \cdot 27)}$$

$$= 1 \cdot 634 \, \text{kN/mm}^2$$

$$= 1634 \, \text{N/mm}^2$$

From these results it is seen that the theoretical compressive strength is given from the 'shear' mode analysis, i.e. the smaller of the two values calculated.

In practice, however, compressive strengths of such magnitude are not achieved. Indeed, such a strength would cause a strain of about 8%, a value far beyond the critical strain of the resin matrix. Due to a number of factors, not the least of which are the voids content, the lack of straightness of the fibres and the variation of G_m with the stress level, the ultimate compressive strengths are found to be very much lower than these analytical results suggest, with failure occurring at a strain of approximately 1·2%. If compressive strengths are calculated on the basis of this ultimate strain value, then from eqns. (5.38) and (5.39)

$$X_c = E_f V_f \times \varepsilon_{x_c}$$

Thus if

$$V_f = 0 \cdot 27,$$
$$X_c = 72 \cdot 4 \times 0 \cdot 27 \times 0 \cdot 012$$
$$= 0 \cdot 235 \, \text{kN/mm}^2$$
$$= 235 \, \text{N/mm}^2$$

In considering the ultimate transverse compressive strength, Y_c, of the composite, this may be reasonably considered to be given approximately by the ultimate compressive strength of the resin matrix. Thus

$$Y_c = 94 \cdot 2 \, \text{N/mm}^2$$

this being the experimentally measured value of the compressive strength of the resin.

The analytical and experimental longitudinal properties obtained for a similar laminate with a value of V_f of 0·64 are shown in Figs. 7.3a and 7.3b for tension and compression respectively, while a comprehensive range of analytical and experimental results for the various unidirectionally reinforced laminae examined is presented in Table 7.1, in which the experimental stiffness properties are the averages of results obtained in tension and compression.

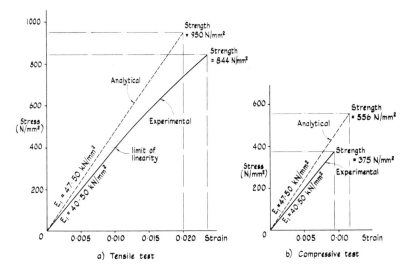

FIG. 7.3. Longitudinal properties of unidirectionally continuously reinforced laminate ($V_f = 0·64$)

7.2.1.2 Properties Obtained from Flexural Tests

Depending upon the composite properties to be investigated, flexural tests may be performed on

(a) long span beams or
(b) short span beams.

Tests to failure on long span beams will ensure failure in a tensile or compressive sense, whereas similar tests on short span beams will cause failure by interlaminar shear.

From long span flexural tests it has been observed that the modulus of elasticity in flexure may be considered the same as that found from tensile or compressive tests, and that the strength obtained is similar to

TABLE 7.1
PROPERTIES OF UNDIRECTIONALLY CONTINUOUSLY REINFORCED LAMINATES
Stiffness properties (tension and compression)

V_f	Density (kg/m^3)	Analytical					Experimental				
		E_1 (kN/mm^2)	E_2 (kN/mm^2)	v_{12}	v_{21}	G_{12} (kN/mm^2)	E_1 (kN/mm^2)	E_2 (kN/mm^2)	v_{12}	v_{21}	G_{12} (kN/mm^2)
0	1·19	—	—	—	—	—	3·22	3·22	0·35	0·35	1·19
0·27	1·57	21·90	4·34	0·31	0·061	1·61	17·68	5·75	0·30	0·098	1·45
0·39	1·73	30·20	5·13	0·29	0·049	1·90	23·82	—	0·28	—	—
0·46	1·83	35·04	5·75	0·28	0·046	2·13	26·99	—	0·28	—	—
0·57	1·98	42·65	7·07	0·26	0·043	2·63	32·79	—	0·28	—	—
0·64	2·07	47·50	8·29	0·25	0·044	3·09	40·50	—	0·27	—	—

Strength properties

V_f	Tensile						Compressive				
	Analytical		Experimental				Analytical		Experimental		
	X_t (N/mm^2)	Y_t (N/mm^2)	X_t (N/mm^2)	Y_t (N/mm^2)	Standard deviation of X_t		X_c (N/mm^2)	Y_c (N/mm^2)	X_c (N/mm^2)	Y_c (N/mm^2)	Standard deviation of X_c
0	—	—	51·5	51·5	7·5		—	—	94·2	94·2	7·1
0·27	438	25·8	429	31·0	5·2		235	94·2	259	88·5	9·2
0·39	604	25·8	568	—	8·1		339	94·2	344	—	10·1
0·46	701	25·8	622	—	5·6		400	94·2	355	—	8·7
0·57	853	25·8	733	—	6·2		495	94·2	392	—	7·6
0·64	950	25·8	844	—	7·3		556	94·2	375	—	8·5

that obtained from a tensile test.[1] At first sight this may appear to contradict the observation that compressive strengths obtained from uniaxial tests tend to be lower than the corresponding tensile strengths. However, if the stress gradient across a section in flexure is considered, it is noticed that the extreme compressive fibres are given stability by the less highly stressed fibres nearer the neutral plane, thus reducing the fibre buckling tendency that occurs in uniform compression and allowing greater strength to be attained.

The interlaminar shearing strength obtained from short span flexural tests has been shown to be given approximately by the shearing strength of the matrix,[1] i.e. by about half the ultimate compressive strength of the matrix, assuming that direct compression produces failure by shear. Thus the *interlaminar shearing strength* for the polyester resin described is given by

$$\tfrac{1}{2} \times 94 \cdot 2$$
$$= 47 \cdot 1 \text{ N/mm}^2$$

The interlaminar shearing strength is little affected by the fibre content but a slight reduction in its value occurs as the fibre content increases, possibly due to the increase in voids that occurs with an increase in fibre volume resulting in a growth of areas of stress concentration.

It can be noted that under flexure, the failure criterion given by eqn. (4.136) reduces to

$$\left(\frac{\sigma_1}{X}\right)^2 + \left(\frac{\tau_{12}}{S}\right)^2 = 1 \qquad (7.3)$$

where in this case σ_1 and X are the longitudinal stress and strength and τ_{12} and S the interlaminar shearing stress and strength respectively.

In long span flexure, τ_{12} is small and hence the second term becomes small in comparison with the first, whereas in short span tests, the relative magnitudes of the two terms are reversed, σ_1 becoming small and the second term predominating.

7.2.1.3 Discussion

The results presented in Table 7.1 show the micromechanical methods developed in Chapter 5 to be quite successful in predicting the stiffness properties of a unidirectionally continuously reinforced laminate, especially the properties E_1 and v_{12}, but to be less consistent when applied to the prediction of the strength characteristics, especially that of X_c, the longitudinal compressive strength. In this latter case, the divergence of analytical and experimental results may be expected in composites with a high fibre content due to the onset of inelastic behaviour in the resin matrix with increase of V_f. Such conclusions may of course be expected when one considers the simplifying assumptions made in the various theoretical derivations.

Figures 7.3a and 7.3b vindicate the assumption that the composite material behaves in a linear manner to failure, although it is noticed that a slight departure from true linearity occurs in the tensile stress–strain curve, probably due to failure of some fibres prior to total rupture.

Flexural tests on short span beams confirm that the interlaminar shearing strength is almost entirely dependent upon the shearing strength of the resin matrix.[1,2]

Although not identical, the order of experimental results reported by other investigators[4-6] tend to be similar to those presented here, this comparison being described in Chapter 9.

7.2.2 Multidirectionally Continuously Reinforced Laminates

Once the stiffness and strength properties of unidirectionally reinforced laminates, and thus the properties of an individual lamina, have been determined, these results may be used to predict the stiffness and strength characteristics of multidirectionally reinforced laminates.

In this section theoretical and experimental investigations for two such laminated composites will be presented, these being of the forms:

(a) four-layered cross-ply laminate, and
(b) two-layered 45° angle-ply laminate.

7.2.2.1 FOUR-LAYERED CROSS-PLY LAMINATE

This laminate consists of four unidirectionally continuously reinforced laminae, each of thickness t, as shown in Fig. 7.4, the fibres of the outer two laminae lying in the x-direction and those of the inner two at right angles to these, i.e. in the y-direction. Because of the orientation of the constituent laminae the inner two laminae are equivalent to a single lamina of thickness $2t$, thus making the composite equivalent to a three-layered symmetric laminate of the form discussed in section 6.2.2.

Each lamina has a fibre volume fraction, V_f, of 0·27. Thus the laminate is constructed from four laminae, each of which is of the form that has been investigated in section 7.2.1.

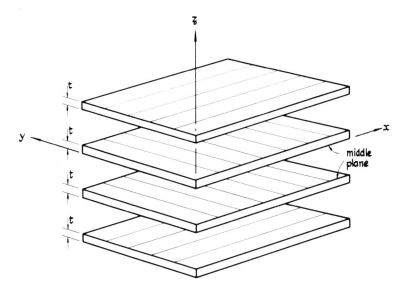

FIG. 7.4. Exploded view of four-layered cross-ply laminate.

(a) *Stiffness Characteristics*

The relationship between the stresses applied to the laminate with respect to the x- and y-directions and the corresponding strains produced can be obtained by using the analytical results developed in Chapter 6, where it has been shown that once the general 3×3 material matrices $[\bar{Q}]$ have been obtained for each constituent lamina, the stiffness characteristics for the lamina can be calculated.

In the case under consideration two such matrices are required, one for the outer laminae and one for the inner ones. Once these are found, then equations similar to eqns. (6.42) and (6.45) may be invoked to obtain the planar and flexural stiffness coefficients of the laminate and from these the equivalent Young's modulus and Poisson's ratio values.

Thus, using the general relationships

$$\begin{bmatrix} \sigma_x \\ \sigma_y \\ \tau_{xy} \end{bmatrix} = [\bar{Q}] \begin{bmatrix} \varepsilon_x \\ \varepsilon_y \\ \gamma_{xy} \end{bmatrix}$$

for the outer layers, denoted by the subscript a,

$$[\bar{Q}]_a = \begin{bmatrix} \dfrac{E_1}{1-v_{12}v_{21}} & \dfrac{v_{21}E_1}{1-v_{12}v_{21}} & \\ \dfrac{v_{12}E_2}{1-v_{12}v_{21}} & \dfrac{E_2}{1-v_{12}v_{21}} & \\ & & G_{12} \end{bmatrix} \qquad (7.4a)$$

and for the inner layers, denoted by the subscript b,

$$[\bar{Q}]_b = \begin{bmatrix} \dfrac{E_2}{1-v_{12}v_{21}} & \dfrac{v_{12}E_2}{1-v_{12}v_{21}} & \\ \dfrac{v_{21}E_1}{1-v_{12}v_{21}} & \dfrac{E_1}{1-v_{12}v_{21}} & \\ & & G_{12} \end{bmatrix} \qquad (7.4b)$$

where, from the example of the unidirectionally reinforced laminate with $V_f = 0.27$ examined previously, the various analytically obtained properties present in these matrices are

$$E_1 = 21.90 \, \text{kN/mm}^2$$
$$E_2 = 4.34 \, \text{kN/mm}^2$$
$$v_{12} = 0.31$$
$$v_{21} = 0.061$$
$$G_{12} = 1.61 \, \text{kN/mm}^2$$

Hence

$$[\bar{Q}]_a = \begin{bmatrix} 22.32 & 1.37 & \\ 1.37 & 4.42 & \\ & & 1.61 \end{bmatrix} \qquad (7.5a)$$

and

$$[\bar{Q}]_b = \begin{bmatrix} 4.42 & 1.37 & \\ 1.37 & 22.32 & \\ & & 1.61 \end{bmatrix} \qquad (7.5b)$$

The in-plane forces per unit width of the laminate are thus given by

$$[N] = \begin{bmatrix} N_x \\ N_y \\ N_{xy} \end{bmatrix} = [\bar{Q}]_a \int_{-2t}^{-t} [\varepsilon] dz + [\bar{Q}]_b \int_{-t}^{t} [\varepsilon] dz + [\bar{Q}]_a \int_{t}^{2t} [\varepsilon] dz$$

$$= ([\bar{Q}]_a 2t + [\bar{Q}]_b 2t)[\varepsilon_0] \tag{7.6}$$

a result analogous to eqn. (6.43). Hence the 'mean' normal and shearing stresses are given by dividing the stiffness terms of this expression by $4t$, the total thickness of the laminate, producing

$$\begin{bmatrix} \sigma_x \\ \sigma_y \\ \tau_{xy} \end{bmatrix} = \tfrac{1}{2}([\bar{Q}_a] + [\bar{Q}]_b) \begin{bmatrix} \dfrac{\partial u_0}{\partial x} \\ \dfrac{\partial v_0}{\partial y} \\ \dfrac{\partial u_0}{\partial y} + \dfrac{\partial v_0}{\partial x} \end{bmatrix} \tag{7.7a}$$

that is

$$\begin{bmatrix} \sigma_x \\ \sigma_y \\ \tau_{xy} \end{bmatrix} = \begin{bmatrix} 13{\cdot}37 & 1{\cdot}37 & \\ 1{\cdot}37 & 13{\cdot}37 & \\ & & 1{\cdot}61 \end{bmatrix} \begin{bmatrix} \dfrac{\partial u_0}{\partial x} \\ \dfrac{\partial v_0}{\partial y} \\ \dfrac{\partial u_0}{\partial y} + \dfrac{\partial v_0}{\partial x} \end{bmatrix} \tag{7.7b}$$

The equivalent Young's modulus and Poisson's ratio values may be obtained by noting that the form of the laminate material matrix implies uncoupled normal and shearing effects, and that the laminate has the same properties in the two orthogonal directions x and y. Denoting Young's modulus and Poisson's ratio in these two directions by E_{lam} and v_{lam}, it may be implied, via eqns. (4.11) or (4.32), that

$$13{\cdot}37 = \frac{E_{\text{lam}}}{1 - v_{\text{lam}}^2} \tag{7.8a}$$

and

$$1.27 = \frac{v_{\text{lam}} E_{\text{lam}}}{1 - v_{\text{lam}}^2} \tag{7.8b}$$

which upon solution gives

$$E_{\text{lam}} = 13.23 \text{ kN/mm}^2$$

and

$$v_{\text{lam}} = 0.102$$

The value of the modulus of rigidity of the laminate, G_{lam}, with respect to the x- and y-directions is still given by the similar property for the individual lamina, that is

$$G_{\text{lam}} = 1.61 \text{ kN/mm}^2$$

a value still independent of E_{lam} and v_{lam}, showing that, even though the laminate possesses identical values of Young's modulus and Poisson's ratio in the two principal directions, it still maintains its orthotropic character and cannot be considered as isotropic.

Although the planar stiffnesses are identical in both the directions x and y, the same will not be found true when considering the stiffnesses in flexure. This is due to the fact that the flexural properties depend on the stacking sequence of the individual laminae.

To illustrate this, consider the calculations of the relationships between the bending moments and torques in the laminate and the corresponding curvatures and twists. The relationships are given by

$$[M] = \begin{bmatrix} M_x \\ M_y \\ M_{xy} \end{bmatrix} = -[\bar{Q}]_a \int_{-2t}^{-t} z[\varepsilon] dz - [\bar{Q}]_b \int_{-t}^{t} z[\varepsilon] dz - [\bar{Q}]_a \int_{t}^{2t} z[\varepsilon] dz$$

$$= \left([\bar{Q}]_a \frac{14}{3} t^3 + [\bar{Q}]_b \frac{2}{3} t^3 \right) [\chi] \tag{7.9}$$

a result similar to eqn. (6.46).

Thus

$$\begin{bmatrix} M_x \\ M_y \\ M_{xy} \end{bmatrix} = \frac{2}{3}t^3 \begin{bmatrix} 160\cdot66 & 10\cdot96 & \\ 10\cdot96 & 53\cdot26 & \\ & & 12\cdot88 \end{bmatrix} \begin{bmatrix} \dfrac{\partial^2 w}{\partial x^2} \\ \dfrac{\partial^2 w}{\partial y^2} \\ 2\dfrac{\partial^2 w}{\partial x \partial y} \end{bmatrix} \quad (7.10)$$

where it is noticed that the diagonal terms in the bending moment equations are different, showing that the greatest stiffness lies in the x-direction and the smallest in the y-direction.

(b) Strength Characteristics

(i) *Tensile strength:* To determine the theoretical tensile strength of the laminate, the process described in section 6.3.1 must be applied.

Considering the laminate subjected to a tensile force in the x-direction of N_x kN/mm, the load–deformation behaviour may be expressed from eqn. (7.6) as

$$\begin{bmatrix} N_x \\ 0 \\ 0 \end{bmatrix} = 2t\,([\bar{Q}]_a + [\bar{Q}]_b)[\varepsilon_0]$$

$$= 2t \begin{bmatrix} 26\cdot74 & 2\cdot74 & \\ 2\cdot74 & 26\cdot74 & \\ & & 3\cdot22 \end{bmatrix} \begin{bmatrix} \dfrac{\partial u_0}{\partial x} \\ \dfrac{\partial v_0}{\partial y} \\ \dfrac{\partial u_0}{\partial y} + \dfrac{\partial v_0}{\partial x} \end{bmatrix} \quad (7.11)$$

from which can be obtained

$$\frac{\partial u_0}{\partial x} = 0.0189 \left(\frac{N_x}{t}\right)$$

$$\frac{\partial v_0}{\partial x} = -0.00194 \left(\frac{N_x}{t}\right) \qquad (7.12)$$

$$\frac{\partial u_0}{\partial y} + \frac{\partial v_0}{\partial x} = 0$$

Hence for the outer laminae

$$\begin{bmatrix} \sigma_x \\ \sigma_y \\ \tau_{xy} \end{bmatrix} = \begin{bmatrix} 22.32 & 1.37 & \\ 1.37 & 4.42 & \\ & & 1.61 \end{bmatrix} \begin{bmatrix} 0.0189 \left(\frac{N_x}{t}\right) \\ -0.00194 \left(\frac{N_x}{t}\right) \\ 0 \end{bmatrix} \qquad (7.13)$$

from which

$$\sigma_x = 0.419 \left(\frac{N_x}{t}\right) \text{kN/mm}^2$$

$$\sigma_y = 0.0173 \left(\frac{N_x}{t}\right) \text{kN/mm}^2 \qquad (7.14)$$

$$\tau_{xy} = 0$$

whilst for the inner laminae

$$\begin{bmatrix} \sigma_x \\ \sigma_y \\ \tau_{xy} \end{bmatrix} = \begin{bmatrix} 4.42 & 1.37 & \\ 1.37 & 22.32 & \\ & & 1.61 \end{bmatrix} \begin{bmatrix} 0.0189 \left(\frac{N_x}{t}\right) \\ -0.00194 \left(\frac{N_x}{t}\right) \\ 0 \end{bmatrix} \qquad (7.15)$$

giving

$$\sigma_x = 0{\cdot}0809 \left(\frac{N_x}{t}\right) \text{kN/mm}^2$$

$$\sigma_y = -0{\cdot}0173 \left(\frac{N_x}{t}\right) \text{kN/mm}^2 \qquad (7.16)$$

$$\tau_{xy} = 0$$

At this stage the equal and opposite character of the values of σ_y in eqns. (7.14) and (7.16) can be noted, a condition which, although compatible with the fact that there is no resultant normal force in the y-direction, does not allow for the fact that at the unloaded edges σ_y must be zero in both laminae. As mentioned in section 6.4, numerical methods are usually applied if a knowledge of laminate behaviour in the vicinity of such edges is required.

In order to determine the value of N_x which will cause failure, the Tsai–Hill failure criterion (eqn. (4.136)) is applied to each lamina. In the case under consideration, because of the absence of shearing stress, this criterion can be reduced to

$$\left(\frac{\sigma_1}{X}\right)^2 + \left(\frac{\sigma_2}{Y}\right)^2 - \frac{\sigma_1 \sigma_2}{X^2} = 1 \qquad (7.17)$$

where X and Y are the lamina strengths corresponding to the direction and sense of the stresses σ_1 and σ_2, i.e. the normal stresses along and perpendicular to the fibre directions, and are given in Table 7.1 for the laminate under investigation.

Application of this criterion to the outer laminae gives

$$\left\{\left(\frac{0{\cdot}419}{0{\cdot}438}\right)^2 + \left(\frac{0{\cdot}0173}{0{\cdot}0258}\right)^2 - \left(\frac{0{\cdot}419 \times 0{\cdot}0173}{0{\cdot}438 \times 0{\cdot}438}\right)\right\} \left(\frac{N_x}{t}\right)^2 = 1$$

from which

$$\frac{N_x}{t} = 0{\cdot}868 \text{ kN/mm}^2$$

whilst application to the inner laminae produces

$$\left\{\left(\frac{-0{\cdot}0173}{-0{\cdot}235}\right)^2 + \left(\frac{0{\cdot}0809}{0{\cdot}0258}\right)^2 - \left(\frac{-0{\cdot}0173 \times 0{\cdot}0809}{-0{\cdot}235 \times -0{\cdot}235}\right)\right\} \left(\frac{N_x}{t}\right)^2 = 1$$

giving

$$\frac{N_x}{t} = 0.318 \text{ kN/mm}^2$$

Thus the inner laminae fail first at a force of

$$N_x = 0.318t \text{ kN/mm}$$

i.e. at an applied stress of

$$\frac{0.318t}{4t} \text{ kN/mm}^2 = 79.5 \text{ N/mm}^2$$

Using eqns. (7.12), at this load, the laminate strain in the x-direction is given by

$$\frac{\partial u_0}{\partial x} = 0.0189 \times 0.318$$

$$= 0.00601$$

and the stresses in the outer laminae just prior to failure of the inner laminae are, from eqns. (7.14),

$$\sigma_x = 0.419 \times 0.318 = 0.133 \text{ kN/mm}^2$$
$$\sigma_y = 0.0173 \times 0.318 = 0.00550 \text{ kN/mm}^2 \qquad (7.18)$$
$$\tau_{xy} = 0$$

It is now assumed that only the outer laminae can carry load, and hence the load–deformation behaviour of the laminate will be expressed as

$$\begin{bmatrix} N_x \\ 0 \\ 0 \end{bmatrix} = 2t[\bar{Q}]_a [\varepsilon_0]$$

$$= 2t \begin{bmatrix} 22.32 & 1.37 & \\ 1.37 & 4.42 & \\ & & 1.61 \end{bmatrix} \begin{bmatrix} \dfrac{\partial u_0}{\partial x} \\ \dfrac{\partial v_0}{\partial y} \\ \dfrac{\partial u_0}{\partial y} + \dfrac{\partial v_0}{\partial x} \end{bmatrix} \qquad (7.19)$$

from which it can be deduced that

$$\frac{\partial u_0}{\partial x} = 0.0228 \left(\frac{N_x}{t}\right)$$

$$\frac{\partial v_0}{\partial y} = -0.00708 \left(\frac{N_x}{t}\right) \quad (7.20)$$

$$\frac{\partial u_0}{\partial y} + \frac{\partial v_0}{\partial x} = 0$$

Hence for the outer laminae

$$\begin{bmatrix} \sigma_x \\ \sigma_y \\ \tau_{xy} \end{bmatrix} = \begin{bmatrix} 22.32 & 1.37 & \\ 1.37 & 4.42 & \\ & & 1.61 \end{bmatrix} \begin{bmatrix} 0.0228 \left(\frac{N_x}{t}\right) \\ -0.00708 \left(\frac{N_x}{t}\right) \\ 0 \end{bmatrix} \quad (7.21)$$

from which

$$\sigma_x = 0.5 \left(\frac{N_x}{t}\right) \text{kN/mm}^2$$

$$\sigma_y = 0 \quad (7.22)$$

$$\tau_{xy} = 0$$

Thus at failure of the inner laminae the stress, σ_x, in the outer laminae increases suddenly from $0.419\,(N_x/t)$ to $0.5\,(N_x/t)$, i.e. from $0.133\,\text{kN/mm}^2$ (eqns. (7.18)), i.e. $133\,\text{N/mm}^2$, to

$$0.5 \times 0.318 = 0.159 \text{ kN/mm}^2$$
$$= 159 \text{ N/mm}^2$$

this sudden increase in stress of $0.026\,\text{kN/mm}^2$ producing, from eqns. (7.20) and (7.22), an increase in strain of $0.0228 \times 0.026/0.5 = 0.00119$.

Since σ_y is now zero, the failure criterion simply reduces to

$$\left(\frac{\sigma_1}{X}\right)^2 = 1 \quad (7.23)$$

i.e. the stress σ_1 in the outer laminae (identical to σ_x) may be increased to the outer laminae tensile stress of 438 N/mm².

Therefore the outer laminae can sustain a further increase of stress equal to $(438-159)$ N/mm², i.e. 0·279 kN/mm² which from eqns. (7.22) implies that N_x can be increased by an amount given by

$$N_x = \frac{0 \cdot 279}{0 \cdot 5} t = 0 \cdot 558 t \text{ kN/mm},$$

and this additional force produces a further strain in the x-direction given from eqns. (7.20) as

$$\frac{\partial u_0}{\partial x} = 0 \cdot 0228 \times 0 \cdot 558 = 0 \cdot 01272$$

Thus at total failure of the laminate the value of the tensile force, N_x, is

$$(0 \cdot 318 + 0 \cdot 558)t = 0 \cdot 876 t \text{ kN/mm}$$
$$= 876 t \text{ N/mm}$$

giving an equivalent ultimate tensile strength, X_t, of $876t/4t$ N/mm², or

$$X_t = 219 \text{N/mm}^2$$

i.e. half the strength of a similar laminate having the fibres of all four laminae lying along the x-direction.

The ultimate tensile strain in the x-direction is given by addition of the three strain components calculated, i.e.

$$0 \cdot 00601 + 0 \cdot 00119 + 0 \cdot 01272 \simeq 0 \cdot 020$$

The analytical results obtained here are compared in Fig. 7.5a with the experimental results obtained from the testing of such a cross-ply laminate, where it is noticed that, although the theoretical predictions appear generally conservative, the postulated form of progressive failure is clearly vindicated.

(*ii*) *Compressive strength*: A similar procedure may be adopted to determine the compressive strength of the laminate. Since the stiffness properties of the constituent laminae are considered the same in both compression and tension, then up to failure of the first lamina, the stresses and strains in the laminate will be the same as in tension but of opposite sign.

To determine the load at which the initial lamina failure occurs, the

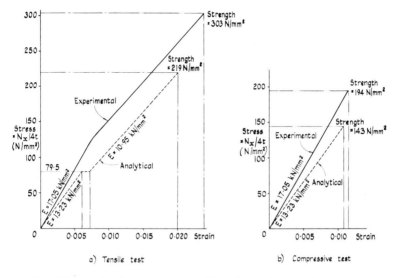

FIG. 7.5. Longitudinal properties of four-layered cross-ply laminate.

Tsai–Hill failure criterion is applied to the individual laminae. Thus for the outer laminae, the criterion produces

$$\left\{\left(\frac{-0{\cdot}419}{-0{\cdot}235}\right)^2+\left(\frac{-0{\cdot}0173}{-0{\cdot}0942}\right)^2-\left(\frac{-0{\cdot}419\times-0{\cdot}0173}{-0{\cdot}235\times-0{\cdot}235}\right)\right\}\left(\frac{N_x}{t}\right)^2=1$$

giving

$$\frac{N_x}{t}=-0{\cdot}570\,\text{kN/mm}^2$$

while for the inner laminae it gives

$$\left\{\left(\frac{0{\cdot}0173}{0{\cdot}438}\right)^2+\left(\frac{-0{\cdot}0809}{-0{\cdot}0942}\right)^2-\left(\frac{0{\cdot}0173\times-0{\cdot}0809}{0{\cdot}438\times0{\cdot}438}\right)\right\}\left(\frac{N_x}{t}\right)^2=1$$

from which

$$\frac{N_x}{t}=-1{\cdot}157\,\text{kN/mm}^2$$

the negative root being taken due to the compressive nature of the load.
Hence in this case it is the outer laminae which are the first to fail, this

occurring when the load per unit width reaches

$$N_x = -0.570t \text{ kN/mm}$$

At this load, the strain in the laminate in the x-direction, again obtained using eqns. (7.12), is given by

$$\frac{\partial u_0}{\partial x} = 0.0189 \times -0.570$$
$$\simeq -0.011$$

and the stresses in the inner laminae at incipient failure of the outer laminae are thus, from eqns. (7.16),

$$\begin{aligned} \sigma_x &= 0.0809 \times -0.570 = -0.0461 \text{ kN/mm}^2 \\ \sigma_y &= -0.0173 \times -0.570 = 0.00986 \text{ kN/mm}^2 \\ \tau_{xy} &= 0 \end{aligned} \qquad (7.24)$$

After failure of the outer laminae, any further load increase must be carried by the inner laminae alone, and hence the load–deformation behaviour of the laminate becomes

$$\begin{bmatrix} N_x \\ 0 \\ 0 \end{bmatrix} = 2t[\bar{Q}]_b[\varepsilon_0]$$

$$= 2t \begin{bmatrix} 4.42 & 1.37 & \\ 1.37 & 22.32 & \\ & & 1.61 \end{bmatrix} \begin{bmatrix} \dfrac{\partial u_0}{\partial x} \\ \dfrac{\partial v_0}{\partial y} \\ \dfrac{\partial u_0}{\partial y} + \dfrac{\partial v_0}{\partial x} \end{bmatrix} \qquad (7.25)$$

from which

$$\frac{\partial u_0}{\partial x} = 0.115 \left(\frac{N_x}{t} \right)$$

$$\frac{\partial v_0}{\partial y} = -0.00708 \left(\frac{N_x}{t} \right)$$

$$\frac{\partial u_0}{\partial y} + \frac{\partial v_0}{\partial x} = 0$$

Thus for the inner laminae

$$\begin{bmatrix} \sigma_x \\ \sigma_y \\ \tau_{xy} \end{bmatrix} = \begin{bmatrix} 4\cdot 42 & 1\cdot 37 & \\ 1\cdot 37 & 22\cdot 32 & \\ & & 1\cdot 61 \end{bmatrix} \begin{bmatrix} 0\cdot 115\left(\dfrac{N_x}{t}\right) \\ -0\cdot 00708\left(\dfrac{N_x}{t}\right) \\ 0 \end{bmatrix} \quad (7.26)$$

from which

$$\sigma_x = 0\cdot 5\left(\frac{N_x}{t}\right) \text{kN/mm}^2$$

$$\sigma_y = 0 \qquad (7.27)$$

$$\tau_{xy} = 0$$

Hence at failure of the outer laminae, the stress, σ_x, in the inner laminae suddenly increases from $0\cdot 0809\,(N_x/t)$ to $0\cdot 5\,(N_x/t)$, i.e. to

$$0\cdot 5 \times -0\cdot 570 = -0\cdot 285 \text{ kN/mm}^2$$
$$= -285 \text{ N/mm}^2$$

Now the inner laminae possess an ultimate compressive strength in the x-direction of only $94\cdot 2 \text{ N/mm}^2$, which means that they cannot sustain this sudden increase of stress and hence will fail simultaneously with the outer laminae.

Hence total compressive failure of the laminate occurs at a compressive load of magnitude

$$0\cdot 570t \text{ kN/mm}$$

giving an equivalent ultimate compressive strength, X_c, of $570t/4t \text{ N/mm}^2$, i.e.

$$X_c = 143 \text{ N/mm}^2$$

The ultimate compressive strain in the x-direction is thus the strain in the laminate at failure of the outer laminae, i.e. $-0\cdot 011$. The analytical and experimental results for the compressive behaviour of this laminate are compared in Fig. 7.5b, where again it is seen that the predicted form of failure is observed in practice.

7.2.2.2 Two-layered 45° Angle-ply Laminate

This composite, shown in Fig. 7.6, is constructed from two unidirectionally continuously reinforced laminae, each of thickness t, and fibre volume fraction, V_f, of 0·27, the fibre direction of the bottom lamina being inclined at 45° to the x-axis, and the fibre direction of the top lamina at $-45°$ to this axis, using the angular convention depicted in

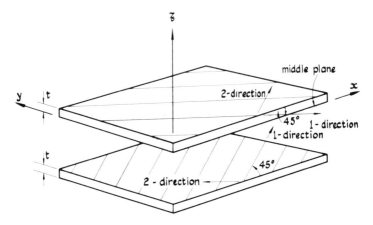

FIG. 7.6. Exploded view of two-layered 45° angle-ply laminate.

Fig. 4.11. Again, the laminate is composed of laminae whose behaviour has been investigated in section 7.2.1, and thus the lamina properties determined previously may be utilised in the description of the laminate.

(a) *Stiffness Characteristics*

The process of constructing the stiffness characteristics of this laminate with respect to the x- and y-axes is similar to that described for the cross-ply laminate, except that the 3×3 material matrices $[\bar{Q}]$ for the laminae are now totally populated due to the coupling that takes place between the normal and shearing strains. The stress–strain relationships for an orthotropic material whose principal material axes are inclined to the axes in which the stresses and strains are required are given from eqns. (4.95) as

$$\begin{bmatrix} \sigma_x \\ \sigma_y \\ \tau_{xy} \end{bmatrix} = \begin{bmatrix} \bar{Q}_{11} \bar{Q}_{12} \bar{Q}_{13} \\ \bar{Q}_{21} \bar{Q}_{22} \bar{Q}_{23} \\ \bar{Q}_{31} \bar{Q}_{32} \bar{Q}_{33} \end{bmatrix} \begin{bmatrix} \varepsilon_x \\ \varepsilon_y \\ \gamma_{xy} \end{bmatrix} = [\bar{Q}] \begin{bmatrix} \varepsilon_x \\ \varepsilon_y \\ \gamma_{xy} \end{bmatrix}$$

where the coefficients $\bar{Q}_{ij}$ in the case of 45° laminae are

$$\bar{Q}_{11} = \bar{Q}_{22} = \frac{(E_1 + E_2 + 2v_{21}E_1)}{4(1 - v_{12}v_{21})} + G_{12}$$

$$\bar{Q}_{12} = \bar{Q}_{21} = \frac{(E_1 + E_2 + 2v_{21}E_1)}{4(1 - v_{12}v_{21})} - G_{12} = \bar{Q}_{11} - 2G_{12}$$

$$\bar{Q}_{33} = \frac{(E_1 + E_2 - 2v_{21}E_1)}{4(1 - v_{12}v_{21})}$$

and

$$\bar{Q}_{13} = \bar{Q}_{31} = \bar{Q}_{23} = \bar{Q}_{32} = \frac{(E_1 - E_2)}{4(1 - v_{12}v_{21})}$$

when $\theta = 45°$, which applies to the bottom lamina, and

$$\bar{Q}_{13} = \bar{Q}_{31} = \bar{Q}_{23} = \bar{Q}_{32} = -\frac{(E_1 - E_2)}{4(1 - v_{12}v_{21})}$$

when $\theta = -45°$, which applies to the top lamina.

Thus, denoting the bottom and top laminae by the suffices a and b respectively, and substituting the numerical values of the principal elastic constants of the laminae, which are given in the description of the four-layered cross-ply laminate, for the bottom lamina

$$[\bar{Q}]_a = \begin{bmatrix} 8.98 & 5.76 & 4.47 \\ 5.76 & 8.98 & 4.47 \\ 4.47 & 4.47 & 6.01 \end{bmatrix} \quad (7.28a)$$

while for the top lamina

$$[\bar{Q}]_b = \begin{bmatrix} 8.98 & 5.76 & -4.47 \\ 5.76 & 8.98 & -4.47 \\ -4.47 & -4.47 & 6.01 \end{bmatrix} \quad (7.28b)$$

Hence the in-plane forces per unit width of the laminate are given by

$$[N] = \begin{bmatrix} N_x \\ N_y \\ N_{xy} \end{bmatrix} = [\bar{Q}]_a \int_{-t}^{0} [\varepsilon] dz + [\bar{Q}]_b \int_{0}^{t} [\varepsilon] dz$$

$$= ([\bar{Q}]_a + [\bar{Q}]_b) t [\varepsilon_0] + ([\bar{Q}]_a - [\bar{Q}]_b) \frac{t^2}{2} [\chi] \quad (7.29)$$

a result analogous to eqn. (6.52), while the out-of-plane moments per unit width are expressed as

$$[M] = \begin{bmatrix} M_x \\ M_y \\ M_{xy} \end{bmatrix} = -[\bar{Q}]_a \int_{-t}^{0} z[\varepsilon]dz - [\bar{Q}]_b \int_{0}^{t} z[\varepsilon]dz$$

$$= ([\bar{Q}]_a - [\bar{Q}]_b)\frac{t^2}{2}[\varepsilon_0] + ([\bar{Q}]_a + [\bar{Q}]_b)\frac{t^3}{3}[\chi] \quad (7.30)$$

a relationship similar to that given by eqn. (6.55).

On substitution of the numerical values given in eqns. (7.28), the force–strain and moment–strain relationships for the laminate are given by

$$\begin{bmatrix} N_x \\ N_y \\ N_{xy} \end{bmatrix} = t \begin{bmatrix} 17 \cdot 96 & 11 \cdot 52 & 0 \\ 11 \cdot 52 & 17 \cdot 96 & 0 \\ 0 & 0 & 12 \cdot 02 \end{bmatrix} \begin{bmatrix} \dfrac{\partial u_0}{\partial x} \\ \dfrac{\partial v_0}{\partial y} \\ \dfrac{\partial u_0}{\partial y} + \dfrac{\partial v_0}{\partial x} \end{bmatrix}$$

$$+ \frac{t^2}{2} \begin{bmatrix} 0 & 0 & 8 \cdot 94 \\ 0 & 0 & 8 \cdot 94 \\ 8 \cdot 94 & 8 \cdot 94 & 0 \end{bmatrix} \begin{bmatrix} \dfrac{\partial^2 w}{\partial x^2} \\ \dfrac{\partial^2 w}{\partial y^2} \\ 2\dfrac{\partial^2 w}{\partial x \partial y} \end{bmatrix}$$

$$\begin{bmatrix} M_x \\ M_y \\ M_{xy} \end{bmatrix} = \frac{t^2}{2} \begin{bmatrix} 0 & 0 & 8 \cdot 94 \\ 0 & 0 & 8 \cdot 94 \\ 8 \cdot 94 & 8 \cdot 94 & 0 \end{bmatrix} \begin{bmatrix} \dfrac{\partial u_0}{\partial x} \\ \dfrac{\partial v_0}{\partial y} \\ \dfrac{\partial u_0}{\partial y} + \dfrac{\partial v_0}{\partial x} \end{bmatrix}$$

$$+ \frac{t^3}{3} \begin{bmatrix} 17\cdot96 & 11\cdot52 & 0 \\ 11\cdot52 & 17\cdot96 & 0 \\ 0 & 0 & 12\cdot02 \end{bmatrix} \begin{bmatrix} \dfrac{\partial^2 w}{\partial x^2} \\ \dfrac{\partial^2 w}{\partial y^2} \\ 2\dfrac{\partial^2 w}{\partial x \partial y} \end{bmatrix}$$

(7.31a)

or more briefly as

$$\begin{bmatrix} N \\ M \end{bmatrix} \begin{bmatrix} A & B \\ B & D \end{bmatrix} \begin{bmatrix} \varepsilon_0 \\ \chi \end{bmatrix}$$ (7.31b)

from which it is seen that in this form of laminate, coupling occurs between the in-plane normal forces and the twist, and between the in-plane shearing forces and the curvatures, these forces and strains being measured relative to the x- and y-axes, with similar coupling occurring between the out-of-plane moments and the in-plane strains.

In order to find the equivalent Young's modulus and Poisson's ratio values for the laminate, consider the application of N_x only to the laminate, the remaining five force and moment values being zero. Such application results in the following values for the strains in terms of N_x obtained from eqns. (7.31a):–

$$\frac{\partial u_0}{\partial x} = 0\cdot1033 \left(\frac{N_x}{t}\right) \qquad \frac{\partial^2 w}{\partial x^2} = 0$$

$$\frac{\partial v_0}{\partial y} = -0\cdot0520 \left(\frac{N_x}{t}\right) \qquad \frac{\partial^2 w}{\partial y^2} = 0 \qquad (7.32)$$

$$\frac{\partial u_0}{\partial y} + \frac{\partial v_0}{\partial x} = 0 \qquad 2\frac{\partial^2 w}{\partial x \partial y} = -0\cdot0572 \left(\frac{N_x}{t^2}\right)$$

showing that twist is produced by the application of normal force.

On dividing N_x by the total thickness of the laminate, $2t$, the mean normal stress σ_x is obtained, and hence the first of eqns. (7.32) may be expressed as

$$\sigma_x = \frac{N_x}{2t} = \frac{1}{2 \times 0.1033} \frac{\partial u_0}{\partial x}$$

$$= 4.84 \frac{\partial u_0}{\partial x}$$

Due to equality of the stiffness properties in the x- and y-directions, on application of N_y only, the similar result

$$\sigma_y = 4.84 \frac{\partial v_0}{\partial y}$$

ensues.

Thus the value of Young's modulus is the same in both directions and is given by

$$E_{\text{lam}} = 4.84 \, \text{kN/mm}^2$$

whilst the single value of Poisson's ratio with respect to the x- and y-directions is given simply by the modulus of the ratio of the lateral to longitudinal strain:

$$\nu_{\text{lam}} = \frac{0.0520}{0.1033} = 0.503$$

The equivalent value of the modulus of rigidity of the laminate with respect to the x- and y-directions is obtained in a similar manner by applying a shearing force N_{xy} only to the laminate. Again using eqns. (7.31a) such application results in the following strain values:

$$\frac{\partial u_0}{\partial x} = 0 \qquad \qquad \frac{\partial^2 w}{\partial x^2} = -0.0572 \left(\frac{N_{xy}}{t^2} \right)$$

$$\frac{\partial v_0}{\partial y} = 0 \qquad \qquad \frac{\partial^2 w}{\partial y^2} = -0.0572 \left(\frac{N_{xy}}{t^2} \right) \qquad (7.33)$$

$$\frac{\partial u_0}{\partial y} + \frac{\partial v_0}{\partial x} = 0.1257 \left(\frac{N_{xy}}{t} \right) \qquad 2 \frac{\partial^2 w}{\partial x \partial y} = 0$$

indicating that curvature is induced by the application of shearing force. Thus, denoting the mean shearing stress, τ_{xy}, by $(N_{xy}/2t)$, the third of eqns. (7.33) gives

$$\tau_{xy} = \frac{N_{xy}}{2t} = \frac{1}{2 \times 0.1257}\left(\frac{\partial u_0}{\partial y} + \frac{\partial v_0}{\partial x}\right)$$
$$= 3.98\left(\frac{\partial u_0}{\partial y} + \frac{\partial v_0}{\partial x}\right)$$

that is,

$$G_{\text{lam}} = 3.98 \text{ kN/mm}^2$$

a value again independent of E_{lam} and v_{lam}.

When considering the effect of bending moments and torques on the laminate, unlike the case of the cross-ply laminate considered previously, the relevant stiffnesses will be identical in both the x- and y-directions due to the material symmetry about these axes. This similarity of stiffness may be noted from the equality of the first two diagonal terms of the $[D]$ matrix of eqns. (7.31).

(b) Strength Characteristics

(i) Tensile strength: Essentially the same procedure is carried out for the determination of the theoretical tensile strengths of angle-ply laminates as for the calculation of those of cross-ply construction.

In the case under consideration, coupling occurs between the planar and flexural displacements, so the effect of a load in the x-direction of N_x kN/mm will be to produce not only in-plane normal strains but also twist. Although this makes the behaviour of this particular composite in some ways more complex than that of the cross-ply laminate previously discussed, the mode of failure will be simpler, since due to the symmetry of each lamina about the direction of loading, failure of both laminae will occur simultaneously.

Therefore, applying a tensile force of N_x kN/mm to the laminate, the load–deformation behaviour may be described using eqns. (7.31) to give the strain values shown in eqns. (7.32).

The twist, $2\partial^2 w/\partial x \partial y$, produces shearing strains in the laminate which, from eqns. (6.19), have the value

$$-z2\frac{\partial^2 w}{\partial x \partial y}$$

implying that the shearing strain varies across the section, being zero at the middle plane and attaining its highest values at the bottom and top faces of the laminate.

Hence on the bottom face, where $z = -t$, the shearing strain is

$$t \times -0.0572\left(\frac{N_x}{t^2}\right) = -0.0572\left(\frac{N_x}{t}\right)$$

whilst on the top face, where $z = t$, it has the value

$$-t \times -0.0572\left(\frac{N_x}{t^2}\right) = 0.0572\left(\frac{N_x}{t}\right)$$

Therefore, using eqns. (7.28a), the stresses on the bottom face are

$$\begin{bmatrix} \sigma_x \\ \sigma_y \\ \tau_{xy} \end{bmatrix} = \begin{bmatrix} 8.98 & 5.76 & 4.47 \\ 5.76 & 8.98 & 4.47 \\ 4.47 & 4.47 & 6.01 \end{bmatrix} \begin{bmatrix} 0.1033\left(\frac{N_x}{t}\right) \\ -0.0520\left(\frac{N_x}{t}\right) \\ -0.0572\left(\frac{N_x}{t}\right) \end{bmatrix} \quad (7.34)$$

this is,

$$\sigma_x = 0.372\left(\frac{N_x}{t}\right) \text{kN/mm}^2$$

$$\sigma_y = -0.128\left(\frac{N_x}{t}\right) \text{kN/mm}^2 \quad (7.35)$$

$$\tau_{xy} = -0.114\left(\frac{N_x}{t}\right) \text{kN/mm}^2$$

Similarly, using eqns. (7.28b), on the top face the stresses are

$$\begin{bmatrix} \sigma_x \\ \sigma_y \\ \tau_{xy} \end{bmatrix} = \begin{bmatrix} 8.98 & 5.76 & -4.47 \\ 5.76 & 8.98 & -4.47 \\ -4.47 & -4.47 & 6.01 \end{bmatrix} \begin{bmatrix} 0.1033\left(\frac{N_x}{t}\right) \\ -0.0520\left(\frac{N_x}{t}\right) \\ 0.0572\left(\frac{N_x}{t}\right) \end{bmatrix} \quad (7.36)$$

giving

$$\sigma_x = 0{\cdot}372\left(\frac{N_x}{t}\right) \text{kN/mm}^2$$

$$\sigma_y = -0{\cdot}128\left(\frac{N_x}{t}\right) \text{kN/mm}^2 \qquad (7.37)$$

$$\tau_{xy} = 0{\cdot}114\left(\frac{N_x}{t}\right) \text{kN/mm}^2$$

At the middle plane the shearing strain becomes zero, and thus in the bottom lamina at this position the stresses are

$$\begin{bmatrix} \sigma_x \\ \sigma_y \\ \tau_{xy} \end{bmatrix} = \begin{bmatrix} 8{\cdot}98 & 5{\cdot}76 & 4{\cdot}47 \\ 5{\cdot}76 & 8{\cdot}98 & 4{\cdot}47 \\ 4{\cdot}47 & 4{\cdot}47 & 6{\cdot}01 \end{bmatrix} \begin{bmatrix} 0{\cdot}1033\left(\frac{N_x}{t}\right) \\ -0{\cdot}0520\left(\frac{N_x}{t}\right) \\ 0 \end{bmatrix} \qquad (7.38)$$

giving

$$\sigma_x = 0{\cdot}628\left(\frac{N_x}{t}\right) \text{kN/mm}^2$$

$$\sigma_y = 0{\cdot}128\left(\frac{N_x}{t}\right) \text{kN/mm}^2 \qquad (7.39)$$

$$\tau_{xy} = 0{\cdot}229\left(\frac{N_x}{t}\right) \text{kN/mm}^2$$

Similarly, in the top lamina the stresses at the middle plane are

$$\sigma_x = 0{\cdot}628\left(\frac{N_x}{t}\right) \text{kN/mm}^2$$

$$\sigma_y = 0{\cdot}128\left(\frac{N_x}{t}\right) \text{kN/mm}^2 \qquad (7.40)$$

$$\tau_{xy} = -0{\cdot}229\left(\frac{N_x}{t}\right) \text{kN/mm}^2$$

The load at which the individual laminae begin to fail may now be deduced by again applying the Tsai–Hill failure criterion to the laminae.

However, since the stresses used in the criterion are defined relative to the principal material axes of the lamina under consideration, these stresses σ_1, σ_2, and τ_{12} must be obtained in terms of the values of σ_x, σ_y, and τ_{xy} given in eqns. (7.35), (7.37), (7.39) and (7.40). These results may easily be obtained by application of the relationship given by eqns. (4.48), where for the bottom lamina $\theta = 45°$ and for the top lamina $\theta = -45°$.

Thus on the bottom face, using eqns. (7.35),

$$\sigma_1 = \left(\frac{1}{\sqrt{2}}\right)^2 \times 0.372\left(\frac{N_x}{t}\right) + \left(\frac{1}{\sqrt{2}}\right)^2 \times -0.128\left(\frac{N_x}{t}\right)$$

$$+ \left(2 \times \frac{1}{\sqrt{2}} \times \frac{1}{\sqrt{2}}\right) \times -0.114\left(\frac{N_x}{t}\right)$$

$$= 0.008\left(\frac{N_x}{t}\right) \text{kN/mm}^2$$

$$\sigma_2 = \left(\frac{1}{\sqrt{2}}\right)^2 \times 0.372\left(\frac{N_x}{t}\right) + \left(\frac{1}{\sqrt{2}}\right)^2 \times -0.128\left(\frac{N_x}{t}\right)$$

$$- \left(2 \times \frac{1}{\sqrt{2}} \times \frac{1}{\sqrt{2}}\right) \times -0.114\left(\frac{N_x}{t}\right) \quad (7.41)$$

$$= 0.236\left(\frac{N_x}{t}\right) \text{kN/mm}^2$$

$$\tau_{12} = -\left(\frac{1}{\sqrt{2}} \times \frac{1}{\sqrt{2}}\right) \times 0.372\left(\frac{N_x}{t}\right) + \left(\frac{1}{\sqrt{2}} \times \frac{1}{\sqrt{2}}\right) \times -0.128\left(\frac{N_x}{t}\right)$$

$$+ \left[\left(\frac{1}{\sqrt{2}}\right)^2 - \left(\frac{1}{\sqrt{2}}\right)^2\right] \times -0.114\left(\frac{N_x}{t}\right)$$

$$= -0.250\left(\frac{N_x}{t}\right) \text{kN/mm}^2$$

On the top face, the stresses relative to the principal material directions are similarly obtained as

$$\sigma_1 = 0.008\left(\frac{N_x}{t}\right) \text{kN/mm}^2$$

$$\sigma_2 = 0.236\left(\frac{N_x}{t}\right) \text{kN/mm}^2 \qquad (7.42)$$

$$\tau_{12} = 0.250\left(\frac{N_x}{t}\right) \text{kN/mm}^2$$

At the middle plane, the corresponding stresses may be similarly obtained, giving for the bottom lamina

$$\sigma_1 = 0.607\left(\frac{N_x}{t}\right) \text{kN/mm}^2$$

$$\sigma_2 = 0.149\left(\frac{N_x}{t}\right) \text{kN/mm}^2 \qquad (7.43)$$

$$\tau_{12} = -0.250\left(\frac{N_x}{t}\right) \text{kN/mm}^2$$

and for the top lamina

$$\sigma_1 = 0.607\left(\frac{N_x}{t}\right) \text{kN/mm}^2$$

$$\sigma_2 = 0.149\left(\frac{N_x}{t}\right) \text{kN/mm}^2 \qquad (7.44)$$

$$\tau_{12} = 0.250\left(\frac{N_x}{t}\right) \text{kN/mm}^2$$

Hence it is noticed that the Tsai–Hill criterion

$$\left(\frac{\sigma_1}{X}\right)^2 + \left(\frac{\sigma_2}{Y}\right)^2 - \frac{\sigma_1\sigma_2}{X^2} + \left(\frac{\tau_{12}}{S}\right)^2 = 1$$

predicts the same value of the failure load for both laminae, this load causing failure at either the outer faces or the middle plane of the laminate.

At the outer faces, the failure criterion gives

$$\left\{\left(\frac{0\cdot008}{0\cdot438}\right)^2+\left(\frac{0\cdot236}{0\cdot0258}\right)^2-\left(\frac{0\cdot008\times0\cdot236}{0\cdot438\times0\cdot438}\right)\right.$$

$$\left.+\left(\frac{0\cdot250}{0\cdot0471}\right)^2\right\}\left(\frac{N_x}{t}\right)^2=1 \tag{7.45}$$

from which

$$\frac{N_x}{t}=0\cdot0946\text{ kN/mm}^2$$

At the middle plane it gives

$$\left\{\left(\frac{0\cdot607}{0\cdot438}\right)^2+\left(\frac{0\cdot149}{0\cdot0258}\right)^2-\left(\frac{0\cdot607\times0\cdot149}{0\cdot438\times0\cdot438}\right)\right.$$

$$\left.+\left(\frac{0\cdot250}{0\cdot0471}\right)^2\right\}\left(\frac{N_x}{t}\right)=1 \tag{7.46}$$

so that

$$\frac{N_x}{t}=0\cdot126\text{ kN/mm}^2$$

It is noticed that in eqns. (7.45) and (7.46), the shearing strength, S, of the laminae is taken as the shearing strength of the matrix, an assumption previously discussed.

Thus laminate failure begins to occur at the outer faces at a load of

$$N_x=0\cdot0946\,t\text{ kN/mm}$$

giving an equivalent ultimate tensile strength, X_t, of $94\cdot6\,t/2t$ N/mm², i.e.

$$X_t=47\cdot3\text{ N/mm}^2$$

The tensile strain in the x-direction at the ultimate tensile load can be obtained from the first of eqns. (7.23) as

$$\frac{\partial u_0}{\partial x} = 0.1033 \times 0.0946$$

$$\simeq 0.010$$

These analytical results are compared in Fig. 7.7a with those obtained from the testing of such a laminate, where it is seen that the form of failure predicted is observed in practice although the analytical results underestimate the measured stiffness and strength values.

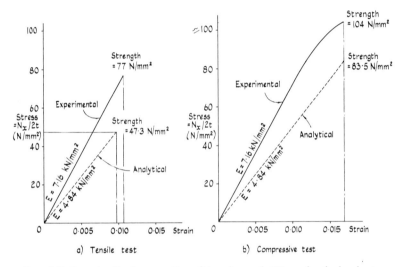

FIG. 7.7. Longitudinal properties of two-layered 45° angle-ply laminate.

(ii) *Compressive strength*: An exactly similar procedure may be used to determine the compressive strength of the laminate.

Considering either the bottom or top lamina, when the load N_x is compressive, application of the Tsai–Hill failure criterion produces, on the outer faces of the laminate,

$$\left\{\left(\frac{-0.008}{-0.235}\right)^2 + \left(\frac{-0.236}{-0.0942}\right)^2 \right.$$

$$\left. - \left(\frac{-0.008 \times -0.236}{-0.235 \times -0.235}\right) + \left(\frac{0.250}{0.0471}\right)^2\right\}\left(\frac{N_x}{t}\right)^2 = 1 \qquad (7.47)$$

giving

$$\frac{N_x}{t} = -0.170 \text{ kN/mm}^2$$

whilst at the middle plane, it gives

$$\left\{\left(\frac{-0.607}{-0.235}\right)^2 + \left(\frac{-0.149}{-0.0942}\right)^2 - \left(\frac{-0.607 \times -0.149}{-0.235 \times -0.235}\right)\right.$$

$$\left. + \left(\frac{0.250}{0.0471}\right)^2\right\}\left(\frac{N_x}{t}\right)^2 = 1 \qquad (7.48)$$

from which is obtained

$$\frac{N_x}{t} = -0.167 \text{ kN/mm}^2$$

Thus when N_x is compressive, laminate failure is initiated at the middle plane at a load of

$$N_x = -0.167 t \text{ kN/mm}$$

giving an equivalent ultimate compressive strength, X_c, of $167t/2t$ N/mm², i.e.

$$X_c = 83.5 \text{ N/mm}^2$$

The compressive strain in the x-direction at this ultimate load is again obtained from the first of eqns. 7.32 as

$$\frac{\partial u_0}{\partial x} = 0.1033 \times -0.67$$

$$\simeq -0.017$$

Figure 7.7b depicts the comparison between these analytical results and the observed experimental values, where again verification of the predicted form of failure is obtained, although the analytical results for both stiffness and strength appear conservative.

Analytical work on the initial failure of both cross-ply and angle-ply laminated composites when subjected to flexure has recently been developed by Turvey,[7-9] the Tsai–Hill criterion having again been utilised as the criterion for failure. In considering the nature of progressive laminate failure, it should be realised that the assumption that on failure a lamina suddenly offers no further contribution to the stiffness and strength of the laminate is a simplification of real laminate behaviour. That this is the case may be seen from the lack of discontinuity in the experimental stress–strain curve shown in Fig. 7.5a. This difference between the analytical and experimental observations has initiated developments of more refined approaches to progressive failure, notably those by Hahn and Tsai.[10]

7.3 DISCONTINUOUSLY REINFORCED LAMINATES

Discontinuous fibres used as the reinforcing medium produce a far narrower spectrum of possible laminates than do continuous fibres, since discontinuous fibres randomly orientated may be considered to produce isotropic laminae, whereas orthotropy is produced with continuous reinforcement.

Thus laminates constructed with such discontinuous reinforcement will, if all the individual laminae are identical, possess the same in-plane isotropic characteristics as the constituent laminae, the stiffness and strength values being dependent upon the glass fibre content.

7.3.1 Stiffness and Strength Properties of Discontinuously Reinforced Laminates

In discontinuously reinforced composites the lengths of the individual fibres usually lie between 6 and 50 mm; it may be noted that within this range there is observed no appreciable variation in composite strength.[11]

The complex nature of discontinuously reinforced composites precludes any accurate analytical techniques for the determination of their stiffness and strength characteristics, so only approximate predictions of these properties may be made. Such values may be obtained from the results of the stiffness and strength analyses of discontinuously reinforced laminae described in Chapter 5.

Consider a laminate reinforced with 6 mm long fibres whose diameters may be considered to be 20×10^{-3} mm. Then using eqn. (5.28) the critical length, l_c, of fibre is

$$l_c = \frac{\sigma_{f\,max} d}{2\tau_y}$$

where $\sigma_{f_{max}}$ is here the ultimate strength of the reinforcement, determined in section 7.2.1 as 1448 N/mm², τ_y is the shearing strength of the matrix, i.e. 47·1 N/mm², and d is the fibre diameter. Thus

$$l_c = \frac{1448 \times (20 \times 10^{-3})}{2 \times 47 \cdot 1}$$

$$= 0 \cdot 307 \text{ mm}$$

which is well below the actual length of the individual fibres.

Hence the ultimate value of the average stress in a 6 mm long fibre is, from eqn. (5.30),

$$\sigma_{f_{av}} = \sigma_{f_{max}}\left(1 - \frac{l_c}{2l}\right)$$

$$= 1448\left(1 - \frac{0 \cdot 307}{2 \times 6}\right)$$

$$= 1411 \text{ N/mm}^2$$

i.e. 97·4% of the tensile stress developed in a continuous fibre.

Now consider such a laminate containing a fibre volume fraction, V_f, of 0·27. If random orientation is assumed, then to determine the stiffness and strength values half the reinforcement may be considered to act in the direction in which these properties are required.

Thus the single value of Young's modulus may be estimated by assuming that the stress in the fibres is 97·4% of that which would occur if the fibres were continuous. This effectively reduces the value of E_f in eqn. (5.6) so that the value of Young's modulus for the composite may be expressed as

$$E = \left([0 \cdot 974 \times 72 \cdot 4] \times \frac{0 \cdot 27}{2}\right) + \left(3 \cdot 22 \times \left[1 - \frac{0 \cdot 27}{2}\right]\right)$$

$$= 9 \cdot 52 + 2 \cdot 79$$

$$= 12 \cdot 31 \text{ kN/mm}^2$$

which is approximately half the stiffness in the fibre direction of a continuously reinforced laminate with the same fibre volume fraction, a result perhaps slightly larger than, although of the order of, that borne out in practice.[11]

The value of Poisson's ratio, v, may be obtained using eqn. (5.11) as

$$v = \left(0.2 \times \frac{0.27}{2}\right) + \left(0.35 \times \left[1 - \frac{0.27}{2}\right]\right)$$

$$= 0.33$$

a result not much different from that obtained for a continuously reinforced laminate, as would be expected due to the similarity in the order of Poisson's ratio for the two basic constituents.

The modulus of rigidity, G, can be obtained from the results for Young's modulus and Poisson's ratio:

$$G = \frac{E}{2(1+v)}$$

$$= \frac{12.31}{2(1+0.33)}$$

$$= 4.63 \text{ kN/mm}^2$$

An estimate of the ultimate tensile strength, X_t, of the laminate may be calculated by application of eqn. (5.34):

$$X_t = \sigma_{fu} V_f + \sigma_{mf} V_m$$

$$= \left([0.974 \times 1.448] \times \frac{0.27}{2}\right) + \left([0.02 \times 3.22] \times \left[1 - \frac{0.27}{2}\right]\right)$$

$$= 0.190 + 0.056$$

$$= 0.246 \text{ kN/mm}^2$$

$$= 246 \text{ N/mm}^2$$

which is rather higher than results obtained by direct experiment.[12] Such results tend to be of the order of a half to two-thirds of the analytical estimate, the higher proportion being attained at high values of fibre volume fraction.[13,14]

In considering the ultimate compressive strength, X_c, of laminates reinforced with randomly orientated discontinuous fibres, experimental evidence suggests that this strength can attain a slightly higher value than the tensile strength, although it is of a similar order.[14] The reason for this relatively high value of compressive strength may be due to the stability against buckling failure given by the random dispersion of the discontinuous fibres.

Some further discussion of the stiffness and strength values of discontinuously reinforced composites is given in Chapter 9.

REFERENCES

1. AL-KHAYATT, Q. J., *The Structural Properties of Glass Fibre Reinforced Plastics*, Ph.D. thesis, University of Aston in Birmingham, 1974.
2. HOLMES, M. and AL-KHAYATT, Q. J., Structural properties of glass reinforced plastics, *Composites*, **6**, July 1975, 157.
3. AINSWORTH, L., The state of filament winding, *Composites*, **2**, July 1971, 14.
4. CLEMENTS, L. L. and MOORE, R. L., Composite properties for E-glass fibres in a room temperature curable epoxy matrix, *Composites*, **9**, April 1978, 93.
5. ROSEN, W. B., Stiffness of fibre composite materials, *Composites*, **4**, Jan. 1973, 16.
6. WAGNER, H. D., FISCHER, S., ROMAN, I. and MAROM, G., The effect of fibre content on the simultaneous determination of Young's and shear moduli in unidirectional composites, *Composites*, **12**, Oct. 1981, 257.
7. TURVEY, G. J., An initial flexural failure analysis of symmetrically laminated cross-ply rectangular plates, *Int. J. Solids and Structure*, **16**, 1980, 451.
8. TURVEY, G. J., Flexural failure of antisymmetric cross-ply plates, *ASCE, J. Eng. Mech. Div.*, Feb. 1981, 279.
9. TURVEY, G. J., Initial flexural failure of square, simply supported, angle-ply plates, *Fibre Science and Technology*, **15**, July 1981, 47.
10. HAHN, H. T. and TSAI, S. W., On the behaviour of composite laminates after initial failure, *J. Composite Materials*, **8**, July 1974, 288.
11. BROUTMAN, L. J. and KROCK, R. H., *Modern Composite Materials*, 1967, Addison–Wesley Publishing, Reading, Mass.
12. BARTON, D. C. and SODEN, P. D., Short-term in-plane stiffness and strength properties of CSM-reinforced polyester laminates, *Composites*, **13**, Jan. 1982, 66.
13. HOLLAWAY, L., *Glass Reinforced Plastics in Construction*, 1978, Surrey University Press, Glasgow.
14. MOLYNEUX, K. W., *An Investigation into the Feasibility of the Structural Use of Glass Reinforced Plastics in Long Span Lightly Loaded Structures*, Ph.D. thesis, University of Aston in Birmingham, 1976.

CHAPTER 8

Time and Temperature Dependent Characteristics of Glass-reinforced Plastics

8.1 INTRODUCTION

The descriptions of glass fibre-reinforced plastics have so far been confined to their stiffness and strength properties obtained from the results of tests carried out over short periods of time, i.e. up to about 3 min, and at ambient temperatures, i.e. in the order of about 10–20°C.

Whilst a knowledge of these properties is of fundamental importance to the understanding of the behaviour of such composite materials, during their service life the loads to which they are subjected, in particular the dead loads, are applied continuously over a period of years. In addition, they may be subjected to a greater or smaller range of temperatures depending upon their environment and upon the use to which they are put.

The behaviour of a number of traditional materials, of which steel is a good example, is very nearly independent of the length of time over which the load is applied (as long as the temperature is not grossly excessive), so that the stiffness and strength values would not show any great variation whether the material were tested over a short period of time or whether long intervals were allowed to elapse between the applications of the increments of load. Also, almost the same values of these structural properties are observed when such tests are carried out over reasonably large temperature ranges,[1] although the magnitude of the strains (or stresses if the strains are suppressed to any degree) would vary due to the effect of the coefficient of expansion.

Other materials, however, are far less resistant to such time and temperature effects, and both thermoplastic and thermosetting resins are among this category, although the thermosetting variety which comprise the polyester and epoxy resins possess far more resistance to such effects than do the thermoplastics.[2]

In glass fibre-reinforced plastics, the glass reinforcement is, like steel, almost impervious to the effects of time and temperature and thus the quantity and form of such reinforcement modifies the behaviour which would occur in the pure plastic, the effects being reduced as the fibre volume fraction increases. Although the adverse effects of time and temperature can be reduced by using suitable quantities of reinforcement and by careful manufacture of the material, nevertheless the effects of time and temperature in glass-reinforced plastics do cause a progressive reduction in both the stiffness and strength of the material and therefore such effects should be considered in the design of these composites.

This chapter is in no way intended to provide a complete review of all the investigations carried out into the dependence of glass fibre-reinforced plastics upon time and temperature, descriptions of some further aspects of such behaviour being found in the references listed at the end of the chapter. Rather it is intended to present a general description of such characteristics from which the effects of various quantities and forms of glass fibre reinforcement may be extrapolated.

8.2 BASICS OF LINEAR VISCOELASTICITY

A material whose stress–strain relationship is dependent upon the length of time over which a stress or strain is applied is known as *viscoelastic*. If, further, for any constant time interval the stress–strain relationship is linear, the material is said to be *linearly viscoelastic*, such behaviour being represented by the equation

$$\sigma = \varepsilon f(t) \tag{8.1}$$

where t is the time.[2] Figure 8.1 shows clearly the nature of the response of such a material.

The term *viscoelastic* is used since such response is characterised by the behaviour of models consisting of elastic springs and viscous dampers, the latter being known as *dashpots*. Similar to the linear proportionality between stress and strain in linear elasticity,

$$\sigma = E\varepsilon \tag{8.2a}$$

in Newtonian viscosity there exists linear proportionality between stress

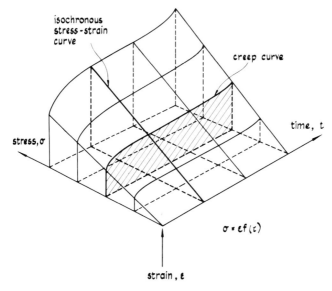

FIG. 8.1. Linear viscoelastic surface at constant temperature.

and rate of change of strain with respect to time, i.e.

$$\sigma = \eta \frac{d\varepsilon}{dt} \quad (8.2b)$$

η being the coefficient of linear viscosity.

The most suitable combinations of these spring and dashpot elements to describe the behaviour of the material may be found by investigating

(a) the mechanism of *creep*, i.e. the deformational behaviour of a material with respect to time when subjected to constant stress, and
(b) the phenomenon of *relaxation*, i.e. the stress decay that occurs in a material under constant strain.

8.2.1 Voigt Model

Consider first the material modelled from a single spring and dashpot connected in parallel as shown in Fig. 8.2a. Such a representation is known as a Voigt model. At any time t, the strain, ε, is the same in both the spring and the dashpot. Thus the stress in the spring, σ_s, is

$$\sigma_s = E\varepsilon \quad (8.3a)$$

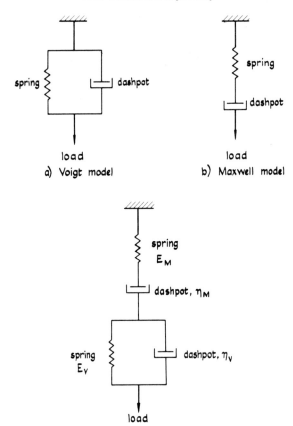

c) Series combination of Voigt and Maxwell models

FIG. 8.2 Linear viscoelastic models.

while the stress in the dashpot, σ_d, is

$$\sigma_d = \eta \frac{d\varepsilon}{dt} \qquad (8.3b)$$

and hence, the equation governing the behaviour of the model is

$$E\varepsilon + \eta \frac{d\varepsilon}{dt} = \sigma_s + \sigma_d = \sigma \qquad (8.4)$$

σ being the total stress in the model.

To represent the creep behaviour, the stress is considered constant and solution of this equation then yields

$$\varepsilon = \frac{\sigma}{E} + \left(\varepsilon_0 - \frac{\sigma}{E}\right)\exp\left(-\frac{E}{\eta}t\right) \quad (8.5a)$$

ε_0 being the initial strain when $t=0$.

The form of curve given by this equation is seen to be determined by the relative values of σ/E and ε_0. Thus to ensure that the strain always increases with time, as would occur in creep behaviour, ε_0 is made equal to zero and hence eqn. (8.5a) is simplified to

$$\varepsilon = \frac{\sigma}{E}\left(1 - \exp\left(-\frac{E}{\eta}t\right)\right) \quad (8.5b)$$

If, after time t, the stress were reduced or removed, then strain recovery would take place, this being described by superimposing on to eqn. (8.5b) a similar equation with a partial or total negative value of stress, the initial value of t being taken as the point at which recovery begins. The form of curve given by eqn. (8.5b), together with the recovery, is shown in Fig. 8.3a.

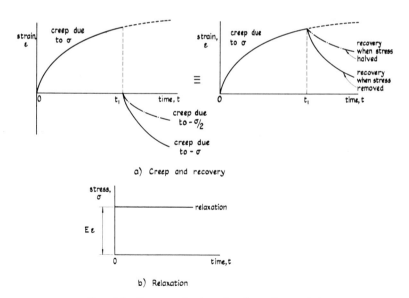

FIG. 8.3. Creep and relaxation in Voigt model.

If on the other hand the strain remains constant, then solution of eqn. (8.4) yields simply

$$E\varepsilon = \sigma \tag{8.6}$$

Hence this model predicts no stress decay at constant strain (Fig. 8.3b), a situation not observed in real viscoelastic materials.

8.2.2 Maxwell Model

Now consider the spring and dashpot connected in series as shown in Fig. 8.2b, this representation being known as a Maxwell model. In this case, at any time t, the stress, σ, in both the spring and dashpot is the same, and hence the behaviour of the two elements is given by

$$\varepsilon_s = \frac{\sigma}{E} \tag{8.7a}$$

and

$$\frac{d\varepsilon_d}{dt} = \frac{\sigma}{\eta} \tag{8.7b}$$

where ε_s and ε_d are the strains in the spring and dashpot respectively.

Differentiating eqn. (8.7a) with respect to time and adding the result to eqn. (8.7b) gives

$$\frac{d\varepsilon_s}{dt} + \frac{d\varepsilon_d}{dt} = \frac{d}{dt}(\varepsilon_s + \varepsilon_d)$$

$$= \frac{d\varepsilon}{dt} = \frac{1}{E}\frac{d\sigma}{dt} + \frac{\sigma}{\eta} \tag{8.8}$$

The creep behaviour of the model is obtained by integration of eqn. (8.8) considering the stress to be constant, giving

$$\varepsilon = \frac{\sigma}{E} + \frac{\sigma}{\eta}t \tag{8.9}$$

where σ/E is the value of the strain when $t=0$, i.e. the initial elastic strain, ε_0.

It can be noted that if the stress, σ, is constant, eqn. (8.8) reduces directly to

$$\frac{d\varepsilon}{dt} = \frac{\sigma}{\eta} \tag{8.10}$$

which on integration and consideration of the initial conditions again produces eqn. (8.9).

Recovery of strain in the Maxwell model is again obtained by superimposition as in the Voigt model, this process being clarified in Fig. 8.4a, which shows both creep and recovery curves obtained from this series arrangement of the components.

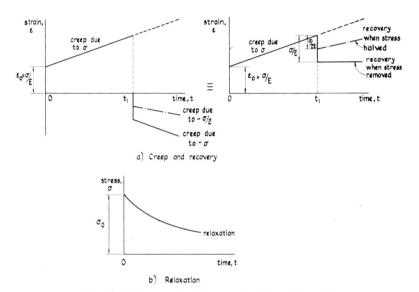

FIG. 8.4. Creep and relaxation in Maxwell model.

The relaxation behaviour of this model is obtained by considering the strain, ε, to remain constant, this constraint allowing eqn. (8.8) to be modified to

$$\frac{1}{E}\frac{d\sigma}{dt}+\frac{\sigma}{\eta}=0 \tag{8.11}$$

from which is obtained

$$\sigma=\sigma_0 \exp\left(-\frac{E}{\eta}t\right) \tag{8.12}$$

where σ_0 represents the initial stress at $t=0$. The curve represented by this equation is shown in Fig. 8.4b, which is a far more realistic simulation of real material behaviour than that expressed by the Voigt model.

8.2.3 Combination of Voigt and Maxwell Models

On inspection of the creep curves given by these two spring and dashpot models, it is noticed that although the curve obtained from the Voigt model is a closer representation of actual creep behaviour than the Maxwell model, it does not represent the initial elastic strain incurred by the initial rapid application of stress, an effect which *is* included in the Maxwell model. Because of these and similar observations relating to the relaxation characteristics, the actual viscoelastic behaviour of a material may be more accurately represented by combining the Voigt and Maxwell models in series, as shown in Fig. 8.2c.[3-5]

In this combined model, the stress in each component is the same and equal to the total stress, while the total combined strain is given by the sum of the two component strains.

Considering the creep behaviour, the Voigt component gives, when subjected to constant stress, σ,

$$\varepsilon_v = \frac{\sigma}{E_v}\left[1 - \exp\left(-\frac{E_v}{\eta_v}t\right)\right] \qquad (8.13a)$$

with the associated description for recovery, while the strain produced by the Maxwell component is given by

$$\varepsilon_m = \frac{\sigma}{E_m} + \frac{\sigma}{\eta_m}t \qquad (8.13b)$$

σ/E_m being the value of ε_m when $t=0$, i.e. the initial elastic strain, ε_0.

On adding the strains ε_v and ε_m in the Voigt and Maxwell models respectively, i.e. superimposing Fig. 8.3a and Fig. 8.4a, creep and recovery curves of the form shown in Fig. 8.5 are obtained, a representa-

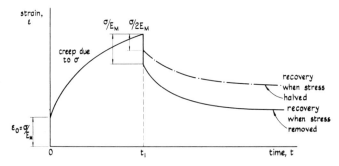

FIG. 8.5. Creep and recovery in Voigt–Maxwell model.

tion much closer to actual creep behaviour than that given by either the Voigt or the Maxwell model alone.

Even greater accuracy may be obtained by modelling the material from a number of Voigt–Maxwell combinations.[3]

These viscoelastic equations apply at a specific temperature, and therefore at any other temperature the various material properties of the models will be changed. The effects of temperature variation are thus treated by a time–temperature superposition.

8.3 DEPENDENCE OF THE STIFFNESS OF GRP COMPOSITES UPON TIME AND TEMPERATURE

A knowledge of the creep characteristics of glass fibre-reinforced plastics is essential in order to estimate the changes in deflection than will occur in a structure with respect to time and with respect to the temperature of its environment.

Intuitively the magnitude of creep in a composite will depend upon the amount and nature of the fibre reinforcement present in the material; independent experimental investigations have shown that this is indeed the case.[6-9]

The form of fibre reinforcement that produces the most highly creep resistant composites is that possessing a continuous unwoven nature, whereas that exhibiting the least resistance consists of short discrete fibres, as exemplified by the chopped strand mat type of reinforcement. Between these two extremes may be placed the bidirectional woven cloth form of reinforcement which shows creep resistant qualities inferior to the unwoven forms probably because of the inherent initial lack of straightness of the fibres arising from its woven nature; under load the fibres tend to straighten out, thus supplementing the creep deflection.

There is experimental evidence to suggest that in some composites the creep in compression is a little less than that in tension,[10] and hence results for tensile creep may tend to produce slightly conservative predictions if applied generally irrespective of the sense of the applied stresses.

Figures 8.6 and 8.7 describe typical results obtained from tensile creep tests at 20°C on composites reinforced with two forms of glass fibre reinforcement, namely chopped strand mat (Fig. 8.6) and unidirectional woven roving, the stress being applied in the direction of the fibres (8.7). The results are plotted on logarithmic scales and are of course the best straight line plots through experimental results which, at times outside

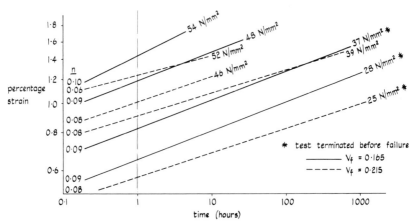

FIG. 8.6. Tensile creep tests on chopped strand mat reinforced composites at 20°C.

the primary stages of creep strain, exhibit approximately linear characteristics.

The idealised linear relationships between the logarithms of the strain and the time suggest that an approximate empirical relationship between these quantities may be written for any particular stress level, thus:

$$\log_{10} \varepsilon = n \log_{10}\left(\frac{t}{t_0}\right) + \log_{10} m \qquad (8.14)$$

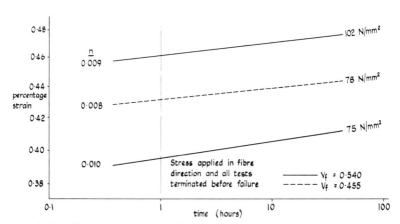

FIG. 8.7. Tensile creep tests on unidirectionally continuously reinforced composites at 20°C.

where ε is the strain, t is the time (h), $t_0 = 1$ h (i.e. unit time), n is the gradient of the curve, and m the value of the strain after 1 h, i.e. when $\log_{10}(t/t_0) = 0$, eqn. (8.14) being applicable to times outside the initial primary creep range.

By taking antilogarithms, eqn. (8.14) may be written in the form

$$\varepsilon = m\left(\frac{t}{t_0}\right)^n \quad (8.15)$$

from which it is seen that the smaller the value of n, the greater the creep resistance of the material. Comparison of the values of n in Figs. 8.6 and 8.7 clearly shows the greater creep resistance of the unidirectional continuously reinforced materials.

It can be noted that eqn. 8.15 gives zero strain when $t = 0$; that is, the initial elastic strain, ε_0, is not accounted for, so it may be considered logical to modify this equation to

$$\varepsilon = \varepsilon_0 + m\left(\frac{t}{t_0}\right)^n \quad (8.16)$$

However, although the initial elastic strain should logically be considered, experimental results have been found to be more appropriate to eqn. (8.15) at times reasonably distant from the origin.[6]

Although these empirical equations are of a slightly different form from those governing the Voigt and Maxwell models previously described, the behaviour of real materials closely resembles that predicted by the combination of the single spring and dashpot models (Fig. 8.5), as is seen from Fig. 8.8 which shows a typical experimentally determined tensile creep and recovery curve plotted on linear scales for a chopped strand mat reinforced composite.[6]

Results such as those presented suggest that over a 50 year period, a unidirectionally continuously reinforced composite containing a reasonably high percentage of fibres may lose approximately 40% of its short term stiffness, whereas a composite containing a fairly low proportion of chopped strand reinforcement may lose in the order of 70% of its initial stiffness value.

In the case of directionally reinforced composites it has been found that when the stress is applied in directions other than along the fibres, an increase in creep behaviour ensues,[6,7] as would be expected because of a reduction in the amount of effective reinforcement in these other directions. This is illustrated in Figs. 8.9 and Fig. 8.10. Fig. 8.9 depicts the

224 GRP in structural engineering

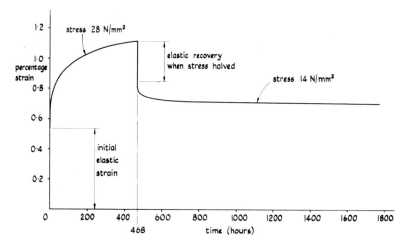

FIG. 8.8. Tensile creep and recovery test on chopped strand mat reinforced composite ($V_f = 0.165$) at 20°C.

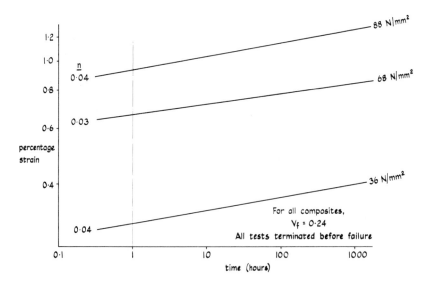

FIG. 8.9. Tensile creep tests on bidirectionally continuously reinforced composites at 20°C (stress applied along a fibre direction).

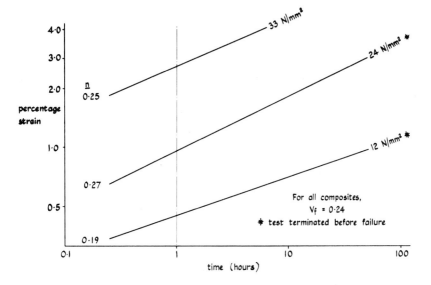

Fig. 8.10. Tensile creep tests on bidirectionally continuously reinforced composites at 20°C (stress applied at 45° to fibre directions).

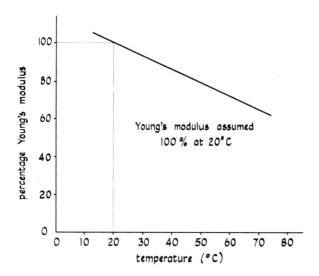

Fig. 8.11. Variation of Young's modulus with temperature.

behaviour of a composite at 20°C reinforced with bidirectional woven roving, both directions containing identical amounts of reinforcement, and stressed along one of the principal material directions. The creep behaviour of the material is seen to be superior to the chopped strand mat form but inferior to the unidirectionally reinforced composites stressed along the fibre directions. On the other hand, the same material, again tested at 20°C but stressed along an axis at 45° to each of the principal material directions, exhibits far inferior creep resistance, as shown in Fig. 8.10. The results for both these tests are again observed to be represented approximately by straight line graphs, indicating that the form of relationships given by eqns. (8.14) and (8.15) are again relevant.

An increase in temperature reduces the stiffness of the material, Fig. 8.11 showing the order of reduction of Young's modulus that may occur due to variations in the environmental temperature.

8.4 DEPENDENCE OF THE STRENGTH OF GRP COMPOSITES UPON TIME AND TEMPERATURE

The strengths of glass fibre-reinforced composites have, like the stiffnesses, been found to reduce with increase in time and environmental temperature. This phenomenon is closely related to the creep mechanism in so much that as the strain increases with time, the weaker bonds in the resin, the weaker fibres, and the weaker fibre-to-resin bonds eventually fail, thus causing higher stresses to be induced in the remaining material. If the applied stresses are of sufficient magnitude this process will eventually lead to failure of the composite.

Because of this relationship between the strength of composites and the creep mechanism, it may be expected that materials with the greatest resistance to creep will show the highest retention of strength, while those forms of glass fibre-reinforced composites showing the greatest propensity to creep will show the greatest reduction in strength with time and temperature. In other words, composites with a high resistance to creep will exhibit both high stiffness and strength retention, and vice versa. There is evidence to suggest that this is indeed the case, composites reinforced with unidirectional continuous fibres showing rather better strength retention characteristics than those reinforced with equivalent quantities of bidirectional woven roving or chopped strand mat.[6,7]

Although many investigations have been conducted into the strength retention characteristics of various forms of glass fibre-reinforced composites with respect to time and temperature, comprehensive results are still lacking, due to the number of parameters involved and to the variability of results. Therefore any definitive statements with regard to such strength characteristics cannot at present be made. However, although the strength retention characteristics of composites will vary with the amount and form of glass fibres present, the reduction in strength with respect to time for such composites may be considered to be of the order shown in Fig. 8.12, the strength reduction under wet conditions being more pronounced than in the dry state. Thus, after a period of 30 years (approximately $2 \cdot 63 \times 10^5$ h) the strength of glass-reinforced composites is of the order of 50% of the short term strength when in the dry condition,[11,12] and only about 35% when in the wet situation,[12] although composites containing a high percentage of unidirectional continuous reinforcement and loaded in the direction of the fibres may show rather more favourable characteristics, whilst those containing a relatively small quantity of chopped strand mat may show a slightly greater reduction in strength.

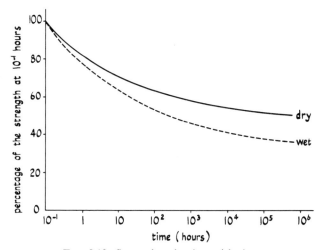

FIG. 8.12. Strength reduction with time.

In the consideration of the effect of temperature on the strength of glass fibre-reinforced plastics, tests on such materials containing 16·5% by volume of chopped strand mat reinforcement and at temperatures of 20°C and 70°C have yielded the results depicted in Fig. 8.13, which show

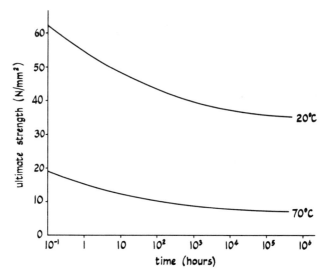

Fig. 8.13. Strength reduction with temperature for chopped strand mat reinforced composite ($V_f = 0.165$).

the substantial reduction in the strength of the material that can be incurred due to a change in temperature of the environment.[6] Composites containing high percentages of unidirectional continuous reinforcement should show a rather smaller amount of such strength degradation, as has been shown by recent investigations.[7]

Therefore, in the design of structures using glass fibre-reinforced plastics, it is imperative that consideration be given to the effect of time and to the effect of the temperature of the environment in which they are to see service, in order that adequate factors of safety in respect of stiffness and strength reductions may be incorporated.

REFERENCES

1. Benham, P. P. and Warnock, F. V., *Mechanics of Solids and Structures*, 1980, Pitman, London.
2. Benjamin, B. S., *Structural Design with Plastics*, 2nd edn. 1982, Van Nostrand Reinhold, New York.
3. Broutman, L. J. and Krock, R. H. *Modern Composite Materials*, 1967, Addison-Wesley Publishing, Reading, Mass.
4. Gillam, E., *Materials under Stress*, 1969, Newnes–Butterworth, London.

5. HOLLAWAY, L., *Glass Reinforced Plastics in Construction*, 1978, Surrey University Press, Glassgow.
6. MOLYNEUX, K. W., *An Investigation into the Feasibility of the Structural Use of Glass Reinforced Plastics in Long Span Lightly Loaded Structures*, Ph.D. thesis, University of Aston in Birmingham, 1976.
7. BHATNAGAR, A. and LAKKAD, S. C., Temperature and orientation dependence of the strength and moduli of glass reinforced plastics, *Fibre Science and Technology*, **14**, 1981, 213.
8. BHATNAGAR, A., LAKKAD, S. C. and RAMESH, C. K., Creep in high glass content unidirectional and bidirectional GRP laminates, *Fibre Science and Technology*, **15**, 1981, 13.
9. JAIN, R. K., GOSWAMY, S. K. and ASTHANA, K. K., A study of the effect of natural weathering on the creep behaviour of glass fibre-reinforced polyester laminates, *Composites*, **10**, Jan. 1979, 39.
10. KABELKA, J., The behaviour of glass reinforced polyester under long term stress, *BFP.*, *5th International Reinforced Plastics Conf.*, Nov. 1966, London.
11. AINSWORTH, L., Properties of glass for plastics reinforcement, *2nd International Reinforced Plastics Conf.*, Nov. 1960, London.
12. BOLLER, K. H., The effect of long-term loading on glass-reinforced plastic laminates, *Proc. 14th Technical and Management Conf., Reinforced Plastics Division SPI*, Feb. 1959, New York.

Part C
DESIGN IN GRP

CHAPTER 9

Properties of GRP Relevant to Structural Design

9.1 INTRODUCTION

Glass-reinforced plastics have been used for some years in structural engineering, although in comparatively small quantities. By far the most common constituent materials used have been 'E' glass fibres, as reinforcement, and polyester resin, as the matrix. These materials have been chosen, rather than others which are available, because they possess economic and structural properties best suited for use in the construction industry. Glass fibres may be used in three different forms, namely chopped strand mat (CSM), woven fabric or rovings.

As has been discussed in previous chapters, the properties of GRP which are important in structural design are the strength and stiffness in tension, compression, flexure and shear, together with a knowledge of how these properties vary with both time and temperature. Numerous theories have been derived for the prediction of these stiffness and strength properties, these having already been discussed in Part B. Since these theoretical predictions are based on a number of assumptions, it is usual to verify them experimentally so that safe design decisions may be made. These experimental results are usually obtained from short term tests. The particular materials used and standards of manufacture can have a considerable effect on the various properties. Where the structure is to have an extended life the structural properties assumed in design must be related to the long, rather than the short, term to ensure that the structure still retains an adequate factor of safety at the end of its life.

It is also important to remember that GRP is an anisotropic material and hence its strength and stiffness properties will be direction dependent. In general the maximum strength and stiffness property will be obtained when the major stress direction is in line with the major fibre direction. Figure 9.1 illustrates the directional tensile strength properties for a GRP laminate with different fibre reinforcement arrangements. The various strength and stiffness properties will now be discussed in more detail.

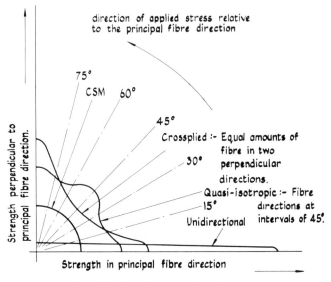

FIG. 9.1. Tensile strength anisotropy.

9.2 SHORT TERM TENSILE AND COMPRESSIVE STRENGTH AND STIFFNESS

The variation of the modulus of elasticity with fibre volume fraction[1-3] is shown in Fig. 9.2a for a GRP material with unidirectional fibres stressed in tension in the fibre direction, where it is seen that experimental results and theoretical predictions are in close agreement. Typical tensile stress–strain relationships for three types of fibre arrangement are shown in Fig. 9.3, together with the order of modulus of elasticity values that they represent. The fibre arrangements shown in Fig. 9.3 are:

(a) unidirectional, with fibre orientation in line with stress direction;
(b) cross-ply, i.e. equal amounts of fibre in two perpendicular directions, with stress direction in line with one of the fibre directions;

(c) CSM (chopped strand mat), giving an approximately isotropic material.

Figures 9.2a and 9.3 relate to tensile modulus. The compressive modulus for most GRP materials is slightly lower than the tensile modulus but for design purposes they are usually considered equal.

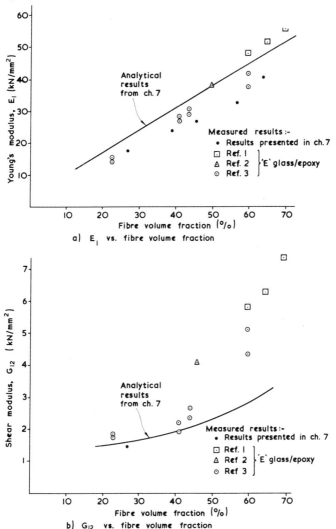

FIG. 9.2. Variation of Young's and shear moduli with fibre volume fraction.

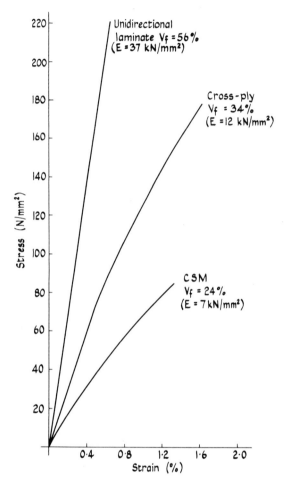

FIG. 9.3. Tensile stress–strain characteristics.

The variation of tensile and compressive strength with fibre volume fraction for unidirectional fibre reinforcement stressed in the fibre direction is shown in Fig. 9.4. For unidirectional reinforcement it is seen that tensile strengths are 50–100% greater than compressive strengths, the greatest difference occuring at high glass fibre contents.

Compressive and tensile properties for cross-ply and CSM for varying fibre volume fractions are also compared in Fig. 9.4. It will be noted that for the cross-ply material the tensile strength is greater than the compre-

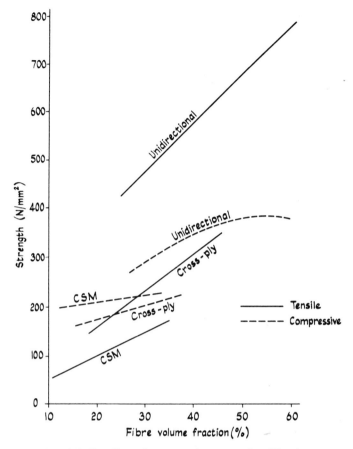

FIG. 9.4. Tensile and compressive strengths of laminates.

ssive strength (as it is for the unidirectional material). However, when CSM reinforcement is used the position is reversed and compressive strength exceeds tensile strength, sometimes by as much as 100%.

9.3 SHORT TERM FLEXURAL STRENGTH AND STIFFNESS

Flexural strength is found commonly to lie between the tensile and compressive strengths of the GRP material. In flexure failure usually occurs in the extreme fibres in compression, but with compressive stresses greater than measured in a pure compression test. This may be due to stress redistribution between extreme fibres (highly stressed) and fibres near the neutral axis of bending (lowly stressed). Such stress redistribution cannot occur in pure compression tests as all fibres are equally stressed.

The flexural modulus may differ from the tensile and compressive moduli particularly at high and low fibre volume fractions, but for design purposes it is generally sufficiently accurate to assume that flexural, tensile and compressive moduli have equal values.

9.4 SHORT TERM SHEARING STRENGTH AND STIFFNESS

The shearing strength of GRP is little affected by fibre volume fraction and has been shown by experiment to be approximately equal to the shearing strength of the resin. A value of $50 \, N/mm^2$ is commonly assumed in design, and is also used in considering interlaminar shearing stresses. Should such stresses exceed the shearing strength, then shear failure between laminations will occur.

The variation of the shear modulus with fibre volume fraction[1-3] for unidirectionally reinforced composites is shown in Fig. 9.2b, where it is noticed that there is a progressive departure of the simple analytical predictions from the experimental results with increase in fibre volume fraction. The shear modulus is seen to vary between about $2 \, kN/mm^2$ for low fibre volume fractions to about $4-7 \, kN/mm^2$ at high values of fibre volume fraction. For cross-ply reinforced composites the shear modulus related to the fibre axes tends to be of the order of half the value for the equivalent unidirectional materials, although the modulus may be a little higher when related to axes inclined at $45°$ to the fibres.

9.5 LONG TERM STRENGTH AND STIFFNESS PROPERTIES OF GRP

Stressed GRP in normal environments exhibits loss of strength with time. This phenomenon has been demonstrated by several research workers,[4-7] and a summary of their findings is shown in Fig. 9.5, from which it is clear that the loss of strength may be significant and must therefore be taken into account when selecting design stresses. Water penetration is generally accepted as a major cause of strength reduction with time,[8-10] and temperature is known to affect the rate of loss of strength.

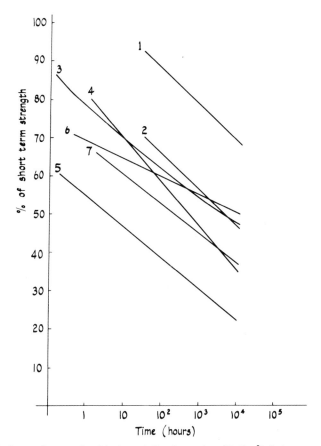

FIG. 9.5. Loss of strength with time. 1, In air tension (Boller[5]); 2, in water tension (Boller[5]); 3, in water tension (Benjamin[4]); 4, in air flexural (Steel[7]); 5, in air flexural (Steel[7]); 6, in air compression (Kabelka[6]); 7, in air tension (Kabelka[7]).

The stiffness of GRP is also time dependent as it is a viscoelastic material which exhibits the phenomenon known as 'creep', i.e. its extension under stress is a function of time. Hence where GRP is stressed for other than very short periods of time, which could be as short as a few minutes at relatively high stress levels, the consideration of only the short term modulus of elasticity is insufficient. Typical results[5,6] are shown in Fig. 9.6 for material in tension or compression. Other work[7] on material subjected to flexure shows (Fig. 9.7) that a third or tertiary stage occurs in flexural creep which is not evident when uniform tensile or compressive stresses are applied.

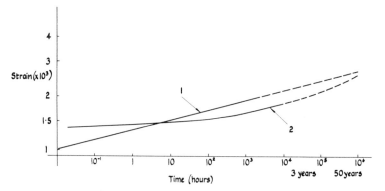

FIG. 9.6. Creep of GRP. 1, Woven fabric, 50% glass by weight, 18°C, 20 N/mm² (Kabelka[6]); 2, woven roving, 23°C, 20 N/mm² (Boller[5]). Polyester resin was used in both cases.

Both resin and fibre contribute to creep but the greater contribution is made by the resin.[11] Thus it is generally found that where short fibres are used, greater creep occurs than where long continuous fibres are employed. At low loadings it has been found that about 95% of the creep strain is recovered after unloading for a period of time of approximately four times the loaded period.

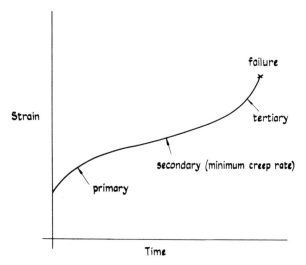

FIG. 9.7. Characteristic creep curve according to Steel[7] (schematic).

Where the strength and stiffness properties of a particular GRP material are known from short term tests, long term properties can be calculated by using various theories[12-14] which have been derived to account for time dependent changes.

9.6 TEMPERATURE EFFECTS ON GRP

The thermal properties of GRP vary with the form and relative quantities of the constituents. Glass reinforced plastic is a better insulator than either steel or aluminium and has a lower coefficient of thermal expansion than aluminium. The thermal properties of GRP are shown in Table 9.1, which also includes the thermal properties of steel and aluminium for comparison.

Increase in temperature, as mentioned in the previous section, also increases the rate of decline in stiffness and strength.

TABLE 9.1
THERMAL PROPERTIES

	Coefficient of expansion ($°C^{-1} \times 10^{-6}$)	Conductivity ($W\,m^{-1}\,°C^{-1}$)
Polyester	9·9–18	0·21
Glass	4·91	1·04
GRP	5–18	0·25–1
Mild steel	11–14	46
Aluminium	22–23	140–190

9.7 PERFORMANCE OF GRP STRUCTURES IN A FIRE

Polyester resins have a low resistance to fire compared with many other structural materials such as steel and concrete. The fire resistance of GRP can be improved, however, by the use of additives such as paraffin wax and chlorinated paraffin. Resins are commercially available that meet the Class 1 requirements in BS476: Part 7:1971, 'Surface spread of flame tests for materials'. This classification requires that a sample of six representative specimens, each 230 mm × 900 mm with a thickness not greater than 50 mm, be such that the flame spread along the surface of the specimens shall not exceed 165 mm at $1\frac{1}{2}$ min, and also shall not have exceeded 165 mm after 10 min, although a tolerance of 25 mm is allowed for one specimen in the sample.

Satisfactory performances of resins can also be obtained when subjected to the test described in BS476: Part 3:1975, 'External fire exposure roof test' for which four specimens, each 1·5 m × 1·2 m, are required. This test is in two parts, the first being a preliminary ignition test and the second a fire penetration and surface ignition test. The preliminary ignition test requires a gas flame 200–250 mm long issuing from an orifice 9·0 ± 0·5 mm in diameter and held 5–10 mm from the surface to be applied to the specimen for 1 min. If penetration occurs during this test no other tests are carried out. Otherwise the remaining three specimens are subjected to the second test in which radiant heat of intensity 14·6 ± 0·5 kW/m^2 is applied for at least 60 min, unless penetration occurs earlier. If desired the test period may be extended to 90 min. In addition the test flame previously described is applied for 1 min at intervals of 5, 10, 15, 30, 45 and 75 min from the start of the test.

The interpretation of these test results is as follows:

Preliminary Ignition Test. The specimen performance is expressed by the letter X if, after withdrawal of the test flame, the duration of flaming exceeds 5 min or if the maximum distance of flaming in any direction is greater than 370 mm. If these responses do not occur then the performance is expressed by the letter P. If fire penetration occurs during the test, the penetration time for the whole sample is given as 1 min.

Fire Penetration and Surface Ignition Test. The fire penetration time for each specimen is recorded to the nearest minute, but if no penetration has occurred the time is taken as the maximum duration of the test.

The extent of surface ignition is given to the nearest 25 mm for the three specimens at 60 min or at the time of penetration.

Thus a classification P45 means that the specimen passed the preliminary ignition test and that penetration occurred after 45 min.

Further details of fire behaviour of plastics is given by Hollaway.[15] The type of fibre reinforcement has also been found to affect performance in fires. Where discontinuous fibres are used they tend to fall away once the resin is burnt off, whereas if continuous woven fibres are used they remain in place after resin burn-off and act as a fire barrier.

9.8 STRUCTURAL JOINTS OR CONNECTIONS

Complete structures normally consist of a collection of structural elements such as beams and columns connected together. The structural connections or joints must be sufficiently strong to transmit the forces from one structural element to the next. Since stress concentrations usually occur in the vicinity of these connections, the design of structural joints requires particular care, especially where the material used is relatively brittle as such materials cannot stress-relieve to the same degree as ductile materials. In comparison with other structural materials, the brittleness of GRP lies between that of concrete and steel. Three basic types of joint are feasible when using glass reinforced plastics, namely adhesive joints, mechanical joints, or joints which use a combination of these effects. Such joints may be made between components of GRP or between components of GRP and those of a different material.

9.8.1 Adhesive Joints

Adhesive joints should be designed so that forces are transferred by shear, compression or tension; peeling effects should be avoided (Fig. 9.8). Adhesives may be used to form three basic types of joint, namely butts, laps and scarfs, which are illustrated in Fig. 9.9. Butt joints afford just a small area of adhesion and are only suitable for resisting compressive forces. Where tensile forces are to be resisted the area of adhesion can be considerably increased by the use of scarf or lap joints. Experiments have shown that scarf joints may give rise to stress concentration factors of 1·5, i.e. the stress at or near the joint may be 1·5 times greater than the axial stress in the connected members. The shearing stress in the adhesive can be controlled by variation of the area of adhesion, as can the interlaminar shearing stress to which the GRP material is subjected in such a joint.

Single lap joints suffer from the disadvantage that the member forces are neither symmetrical nor in line, and consequently stress concentrations tend to be large. Reduced stress concentrations result from the use of double lap joints, due to their symmetry. Stress concentrations are also reduced by bevelling the ends of the connected members. Should resistance to flexure be required, then high peel-strength adhesives should be used.[16]

All adhesive joints suffer from a number of disadvantages. Joint strength is not immediately developed and joints need to be held in

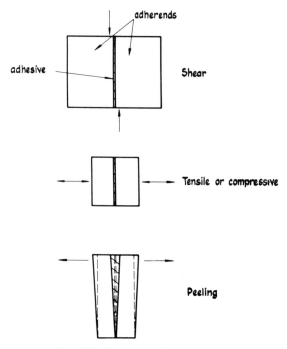

FIG. 9.8. Types of adhesive loading.

position for long periods until the adhesive achieves its full strength. The adhesives themselves are usually polymers and are therefore subject to creep and strength reduction with time. If joints are site made, rather than factory made, quality control may be difficult and inspection after joint completion will not necessarily reveal faults.

In addition, in an adhesive joint made between GRP and metallic components, any change of temperature to which the joint is subjected will cause thermal stresses to be produced due to the different coefficients of linear expansion of the adherents, thus further complicating the design of the joint.

9.8.2 Mechanical Joints
Bolted joints may transfer load from member to member by shear in the bolts or by developing friction between the connected members which is capable of transferring load between them. Transfer of load by friction requires the use of high tension bolts tightened to give high bolt stress.

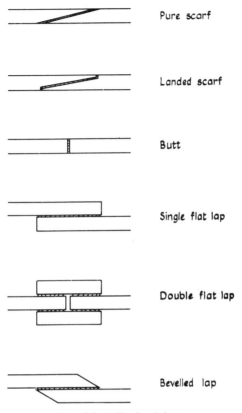

FIG. 9.9. Adhesive joints.

The normal force thus developed between connected members gives considerable frictional resistance. Such friction joints are commonplace in steel structures, but are not suitable for GRP materials. Creep of the GRP will relieve the bolt force and hence reduce the frictional resistance of the joint, and the high bolt forces required tend to cause premature cracking of the GRP material. Bolted joints which rely for load transfer on shear in the bolts are therefore to be preferred. Examples of this type of joint using laps, double laps, scarfs or flanges are shown in Fig. 9.10. Glass fibre reinforcement should be taken round the bolt holes as shown in Fig. 9.11.

The factors to be considered when designing a bolted joint in GRP are similar to those which apply to bolted joints in steel. Unfortunately,

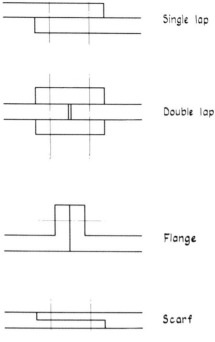

FIG. 9.10. Mechanical joints.

detailed design data applicable to GRP are, as yet, more limited than those available for steel. These limited data[17,18] suggest the following:

Minimum end distance
(centre of bolt hole to end of member)   4·5 D
Minimum edge distance
(centre of bolt hole to edge of member)   3·0 D (woven fabric)
                                                              3·5 D (CSM)

where D = bolt diameter.

With these minimum end and edge distances, tests have shown that failure occurs by crushing of the GRP material in the region where it bears against the bolt. The failure bearing stress is dependent on the compressive strength of the GRP material. Thus for a given bolt shear force (i.e. a given total cross-sectional area of bolt) it is preferable to use several small diameter bolts as this will result in a greater total bearing area. However, care should be exercised when designing such joints as

Properties of GRP relevant to structural design   247

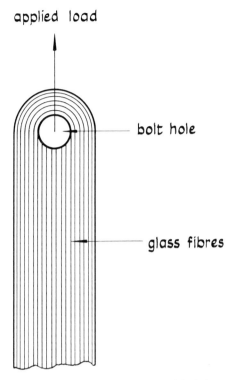

FIG. 9.11. Improved fibre orientation around bolt in a tensile joint.

recent investigations have shown that it is dangerous to extrapolate results for single bolt connections to multi-bolt situations.[19]

9.8.3 Combination Joints

Some possibilities for combination joints are shown in Fig. 9.12. In Fig. 9.12a metal inserts are moulded into the GRP material and subsequently bolted to metal plates to form a double lap joint. Figure 9.12b illustrates a joint in which thin metal strips are cast into the GRP material with their ends protruding. Subsequently the protruding ends are connected by bolting to metal lap plates. The joint of Fig. 9.12c combines bolting and use of adhesive. Apart from the strength they add to the completed joint, the bolts serve to hold the joint in place while the adhesive cures. The bolts are placed towards the ends of the lap to prevent any 'peeling' of the adhesive.

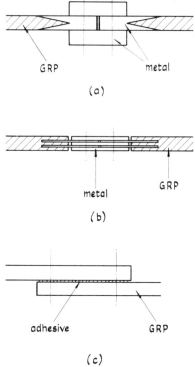

FIG. 9.12. Combination joints.

9.9 STRUCTURAL PROPERTIES OF GRP RELATIVE TO OTHER STRUCTURAL MATERIALS

The influence of the material of construction on the load carrying capacity of a structure depends upon its tensile, compressive and flexural strengths, providing that its mode of failure is not associated with compression buckling. Where buckling is critical the load carrying capacity is dependent upon the modulus of elasticity of the material from which it is constructed. Structural deflections are affected by the modulus of elasticity of the chosen material of construction.

A comparison of the strength and modulus of elasticity of GRP with commonly used structural materials such as steel and concrete give an indication of the relative structural performance. Such a comparison of strengths is given in Fig. 9.13, from which it is seen that some GRP materials have strength properties which are superior to those of structural mild steel. Figure 9.13 also shows a comparison of the specific strengths of construction materials, obtained by dividing the strength of the material by its specific gravity. This comparison shows GRP to be

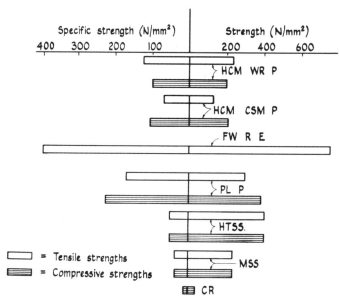

FIG. 9.13. Strengths of structural materials. Key: HCM, hot compression moulding; FW, filament winding; PL, pultrusion; CSM, chopped strand mat; WR, woven rovings; HTSS, high tensile structural steel; MSS, mild structural steel; CR, concrete; R, rovings; P, polyester; E, epoxy.

greatly superior to other structural materials. Specific strengths are particularly relevant where structure self-weight is of a significant magnitude relative to the imposed load on the structure; a more detailed consideration of this point is given later. It should be borne in mind that the strength properties shown in Fig. 9.13 are short term. Where long term strength properties must be used, GRP design stresses would be lower than those shown in Fig. 9.13 but steel and concrete design stresses would remain unchanged.

Where buckling is critical, or for the determination of deflections, structural performance is affected by the modulus of elasticity of the material. The short term modulus of elasticity of GRP, depending on type and amount of glass fibre used, may lie in the range 5–40 kN/mm². Structural steel has a much higher modulus value of 210 kN/mm². However, if specific moduli are compared, the values are approximately 4–25 kN/mm² for GRP and 25 kN/mm² for steel. Once again, if long term performance is to be considered the above values must be reduced for GRP but not for steel. The relatively low value of the modulus of elasticity of GRP can be compensated for by adopting structural forms for GRP which have considerable depth, such as folded plate, shell, arch, stressed skin or deep beam structures.

The importance of some of the factors mentioned above, e.g. ratio of structure self-weight to imposed load and the relevance of specific material properties, will be illustrated by reference to a deep beam or plate structure. The following analysis applies equally well to the I-beam, box-beam or plate structure illustrated in Fig. 9.14. Suppose any of these

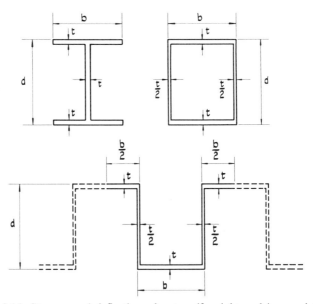

FIG. 9.14. Stresses and deflections due to self weight and imposed load.

structures is required to support its own self-weight plus an imposed load of w per unit run over a span of L.

$$\text{Total load per unit run to be supported} = w + \frac{(2b+d)tL\Delta}{L}$$

Properties of GRP relevant to structural design   251

where $\Delta = \rho g$, ρ being the density of the construction material and g the gravitational acceleration.
The second moment of area of the cross-section, I, is given by:

$$I = \frac{bt^3}{12} + 2bt\left(\frac{d}{2}\right)^2 + \frac{td^3}{12}$$

$$= \frac{bd^3}{12}\left[\left(\frac{t}{d}\right)^3 + 6\frac{t}{d} + \frac{t}{b}\right]$$

Now $(t/d) \ll 1$, and hence $(t/d)^3$ may be neglected with respect to (t/d). Further, if a common practical value of the ratio b/d of $\frac{1}{2}$ is taken

$$I = \frac{td^3}{3}$$

Maximum (mid-span) deflection

$$= \frac{5L^4}{384\,EI}\left[w + (2b+d)t\Delta\right]$$

$$= \frac{15L}{384\,Et}\left(\frac{L}{d}\right)^3 [w + 2dt\Delta] \quad (9.1)$$

Equation (9.1) shows that deflections are markedly affected by the span/depth ratio raised to the third power. Thus a material with a low modulus of elasticity, such as GRP, can be used structurally without incurring unacceptably large deflections, providing the span/depth ratio is kept low.

The relative effect of imposed load and self-weight can best be examined by considering two extreme cases. These are a short span structure with high imposed load and a large span structure with low imposed loading.

In the first case,

$$w \gg 2dt\Delta$$

and hence eqn. (9.1) becomes:

$$\text{Maximum deflection} = \frac{15}{384}\left(\frac{L}{d}\right)^3 \frac{Lw}{t}\frac{1}{E} \quad (9.2)$$

252   GRP in structural engineering

In the second case,

$$w \ll 2dt\Delta$$

and hence eqn. (9.1) becomes:

$$\text{Maximum deflection} = \frac{15}{384}\left(\frac{L}{d}\right)^3 2dL\frac{\Delta}{E} \qquad (9.3)$$

Equation (9.2) shows that, for the short span structure with high imposed load, the deflection depends on modulus E, whereas for the large span structure with low imposed load, eqn. (9.3) shows that the deflection depends on specific modulus E/Δ. Glass-reinforced plastic material thus has its major advantage over other structural materials in the area of long span lightly loaded structures.

It will now be shown that the same conclusion is reached from a consideration of stress levels. Once again the structures shown in Fig. 9.14 are considered.

$$\text{Maximum stress at mid-span} = \sigma = \frac{M}{Z}$$

$$Z = \frac{I}{d/2} = \frac{td^3}{3}\frac{2}{d} = \frac{2td^2}{3}$$

$$M = \frac{L^2}{8}(w + 2dt\Delta)$$

Hence

$$\text{Maximum stress at mid-span} = \sigma = \frac{3}{16t}\left(\frac{L}{d}\right)^2 (w + 2dt\Delta)$$

Let the permissible stress be σ_p; then

$$\frac{\sigma}{\sigma_p} = \frac{3}{16t\sigma_p}\left(\frac{L}{d}\right)^2 (w + 2dt\Delta)$$

In design the maximum permitted value of σ/σ_p is unity; hence

$$\frac{3}{16t\sigma_p}\left(\frac{L}{d}\right)^2 (w + 2dt\Delta) \leq 1$$

If the same two extreme cases are considered; i.e. $w \gg 2dt\Delta$ and $w \ll 2dt\Delta$, then the former gives:

$$\frac{3}{16}\left(\frac{L}{d}\right)^2 \frac{w}{t} \frac{1}{\sigma_p} \leq 1 \qquad (9.4)$$

and the latter gives:

$$\frac{3}{16}\left(\frac{L}{d}\right)^2 2d \frac{\Delta}{\sigma_p} \leq 1 \qquad (9.5)$$

Again it is seen that the span/depth ratio is important in controlling stress levels as it occurs in both eqns. (9.4) and (9.5) raised to the second power. Equation (9.4) shows that, for a short span structure with high imposed load, load capacity depends on strength (σ_p), whereas for a large span structure with low imposed load, eqn. (9.5) shows that the load capacity depends on specific strength (σ_p/Δ). These conclusions, obtained from consideration of stress levels, are the same as those derived earlier from a consideration of deflection.

9.10 TRANSFORMED SECTIONS

Where a structural member is composed of two or more materials which have differing values of modulus of elasticity, it is advantageous to transform the actual cross-section into an equivalent cross-section of a single material. This technique has been used for some time for reinforced concrete structures (steel has a modulus of elasticity about 15 times greater than concrete) and may equally be applied to GRP laminates.

Consider a GRP material consisting of two layers, one of thickness t_a with modulus of elasticity E_a and the other of thickness t_b and modulus of elasticity E_b, subjected to an axial force P as shown in Fig. 9.15a. The layers are of breadth b. Let the strain in the two layers be ε; then the stresses in the two layers are given by:

$$\sigma_a = \varepsilon E_a$$
$$\sigma_b = \varepsilon E_b$$

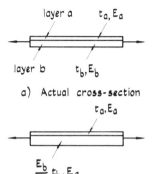

FIG. 9.15. Transformed section (axial force).

and the total force in the cross-section is:

$$P = \sigma_a b t_a + \sigma_b b t_b$$
$$= \varepsilon E_a b t_a + \varepsilon E_b b t_b$$
$$= \varepsilon E_a b \left(t_a + \frac{E_b}{E_a} t_b \right)$$

Therefore,

$$\frac{P}{b\left(t_a + \dfrac{E_b}{E_a} t_b\right)} = \varepsilon E_a \qquad (9.6)$$

Now consider the transformed section (Fig. 9.15b) in which all the material has a modulus of elasticity of E_a. This transformed section is obtained by multiplying the thickness of layer b by E_b/E_a so that its transformed thickness is $t_b(E_b/E_a)$. The following equation may then be written for the transformed section:

$$\sigma = \varepsilon E$$

i.e.

$$\frac{P}{b\left(t_a + \dfrac{E_b}{E_a} t_b\right)} = \varepsilon E_a \qquad (9.7)$$

It will be seen that eqns. (9.6) and (9.7) are identical, and hence the transformed section may be used as a substitute for the actual cross-section. It must be noted that the actual stress in layer a is equal to the stress in the transformed section, i.e.:

$$\sigma_a = \frac{P}{b\left(t_a + \frac{E_b}{E_a} t_b\right)}$$

whilst the actual stress in layer b is equal to the stress in the transformed section multiplied by E_b/E_a, i.e.:

$$\sigma_b = \sigma_a \frac{E_b}{E_a} = \frac{P}{b\left(t_a + \frac{E_b}{E_a} t_b\right)} \cdot \frac{E_b}{E_a}$$

Transformed sections may also be used for cross-sections subjected to bending moments. Consider the T-section composed of two materials (elements a and b) and its transformed equivalent composed entirely of element a material (Fig. 9.16). Both sections will have the same bending properties providing that the resultant normal forces on the cross-sections are the same at equal strain levels. Providing that the thickness of element b is transformed to $t_b(E_b/E_a)$, the resultant normal forces will be the same, as can be seen by comparing Figs. 9.16 d and h, thus:

$$P_a = \frac{\varepsilon_a E_a + \varepsilon_c E_a}{2} \cdot b t_a$$

$$P_b = \frac{\varepsilon_c E_b}{2} t_b (n - t_a) = \frac{\varepsilon_c E_a}{2} t_b \frac{E_b}{E_a} (n - t_a)$$

$$P_c = \frac{\varepsilon_b E_b}{2} \cdot t_b (d - n) = \frac{\varepsilon_b E_a}{2} \cdot t_b \frac{E_b}{E_a} (d - n)$$

It is, therefore, the case that the theory of bending may be applied to the transformed section to calculate such quantities as depth of neutral axis, second moment of area, bending stress and deflection. Such calculated values will give the performance of the actual beam. It must be remembered, however, that stresses calculated for the transformed section

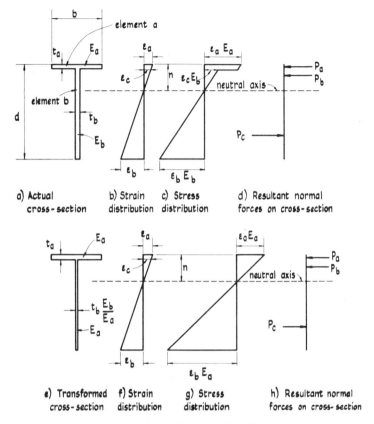

FIG. 9.16. Transformed section (bending moment).

must be modified to give the actual stresses in element 2. Thus, referring to Figs. 9.16c, g:

Stress in transformed section, lower extreme fibre $= \dfrac{M(d-n)}{I} = \varepsilon_b E_a$

Stress in actual section, lower extreme fibre $= \dfrac{M(d-n)}{I} \cdot \dfrac{E_b}{E_a} = \varepsilon_b E_b$

From the above argument it is clear that the procedures for the use of transformed sections are the same for applied axial force as they are for applied bending moment.

REFERENCES

1. CLEMENTS, L. L. and MOORE, R. L., Composite properties for E-glass fibres in a room temperature curable epoxy matrix, *Composites*, **9**, April 1978, 93.
2. ROSEN, W. B., Stiffness of fibre composite materials, *Composites*, **4**, Jan. 1973, 16.
3. WAGNER, H. D., FISCHER, S., ROMAN, I. and MAROM, G., The effect of fibre content on the simultaneous determination of Young's and shear moduli in unidirectional composites, *Composites*, **12**, Oct. 1981, 257.
4. BENJAMIN, B. S., *Structural Design with Plastics*, 2nd edn., 1982, Van Nostrand Reinhold, New York.
5. BOLLER, K. H., The effect of long-term loading on glass-reinforced plastic laminates, *Proc. 14th Technical and Management Conf., Reinforced Plastics Division SPI*, Feb. 1959.
6. KABELKA, J., The behaviour of glass reinforced polyester under long term stress, *BFP 5th International Reinforced Plastics Conf.*, London, Nov. 1966.
7. STEEL, D. J., The creep and stress rupture of reinforced plastics, *Trans. J. Plastics Inst.*, Oct. 1965, 161.
8. GARG, A. C. and TROTMAN, C. K., Influence of water on fracture behaviour of random fibre glass composites, *Engineering Fracture Mechanics*, **13**, 1980, 357.
9. HULL, D., *An introduction to composite materials*, 1981, Cambridge University Press, Cambridge, UK.
10. PRITCHARD, G., Environmental degradation of hybrid composites, in *Fibre Composite Hybrid Materials*, ed. N. L. Hancox, 1981, Applied Science Publishers, London.
11. BHATNAGAR, A., LAKKAD, S. C. and RAMESH, C. K., Creep in high glass content unidirectional and bidirectional GRP laminates, *Fibre Science and Technology*, **15**, 1981, 13.
12. MCLAUGHLIN, J. R., A new creep law for plastics, *Modern Plastics*, Feb. 1968, 97.
13. LARSON, F. R. and MILLER, J., A time temperature relationship for rupture and creep stresses, *Trans. ASME*, **74**, 1952, 765.
14. GOLDFEIN, S., Time, temperature and rupture stresses in reinforced plastics, *Modern Plastics*, Dec. 1954, 148.
15. HOLLAWAY, L., *Glass reinforced plastics in construction*, 1978, Surrey University Press, Glasgow.
16. ALLRED, R. E. and GUESS, T. R., Efficiency of double-lapped composite joints in bending, *Composites*, **9**, April 1978, 112.
17. STRAUSS, E. L., Effects of stress concentration on the strength of reinforced plastic laminates, *Proc. 14th Technical and Management Conf., Reinforced Plastics Division SPI*, Feb. 1959, New York.
18. WEISS, M. D., Mechanical fasteners for glass reinforced plastics, *Proc. 14th Technical and Management Conf., Reinforced Plastics Division SPI*, Feb. 1959, New York.
19. PYNER, G. R. and MATTHEWS, F. L., Comparison of single and multi-hole bolted joints in glass fibre reinforced plastic, *J. Composite Materials*, **13**, July 1979, 232.

CHAPTER 10

Design of GRP Box-beams

10.1 INTRODUCTION

The relatively low elastic moduli of GRP necessitates its structural use in forms where adequate stiffness can be attained by virtue of the structural shape, and hence the material finds its predominant utilisation in closed forms such as pipes, and in folded plate, shell and stressed skin structures.[1-3]

In this chapter structural design in GRP is illustrated with reference to the design of a closed form of structure, namely the box-beam. This structural form has been chosen as a design example as it is one of the most common and efficient forms of utilising fibre reinforced materials in structural engineering. Moreover, box-beam design demonstrates the manner in which different types of fibre reinforcement may be selected to be most appropriate to the function to be served. Thus different GRP materials may be selected for compression flange, tension flange and web.

Because of the efficiency of the box-beam form in GRP, studies into various aspects of its structural behaviour in the medium have been carried out by a number of investigators[4-7] and the interested reader is encouraged to become acquainted with the results obtained. In this chapter are presented the typical design of box-beams in GRP and the experimental testing carried out on the designed sections under both short and long term loading[8,9] so that a comparison can be made between the actual and predicted behaviour.

10.2 LOADING, SPAN AND CROSS-SECTIONAL SHAPE

The beams were designed to support a short term design load of 25 kN over a span of 3 m, the design load to be equally divided between third span points and the design to give a factor of safety of two.

The proportions to be used in the beam cross-section were decided upon by consideration of the flexural and torsional stiffnesses (EI and GJ) and the compression buckling characteristics.

For the cross-section shown in Fig. 10.1, the second moment of area is:

$$I = (b+t)t\left(\frac{d-t}{2}\right)^2 + \frac{2bt^3}{12} + \frac{2t(d-t)^3}{12} \qquad (10.1)$$

and as the section is thin walled terms in t^2 and t^3 can be neglected so that this expression may be reduced to:

$$I = 2bt\left(\frac{d}{2}\right)^2 + \frac{2td^3}{12} \qquad (10.2)$$

The torsional second moment of area is equal to:

$$J = \frac{4(bd)^2}{\frac{2}{t}(b+d)}$$

$$= \frac{2tb^2d^2}{b+d} \qquad (10.3)$$

A suitable size of beam cross-section for the specified loading and span is given by the relationship:

$$b + d = 450 \text{ mm} \qquad (10.4)$$

then eqns. (10.2), (10.3) and (10.4) may be represented graphically as in Fig. 10.1. From Fig. 10.1 a cross-sectional shape may be selected from which values for both flexural and torsional stiffness (I and J) may be read. The shape selected for design in this case is defined by $d = 2b$. Thus:

$$b = 150 \text{ mm}$$
$$d = 300 \text{ mm}$$
$$I = 11\cdot25 \times 10^6 \, t \text{ mm}^4$$
$$J = 9\cdot00 \times 10^6 \, t \text{ mm}^4$$

This initial selection of cross-sectional shape assumes that the wall thickness t is constant. Subsequently it will be shown that there is an advantage in making the compression flange thicker than the remainder of the cross-section to reduce the tendency for local compression buckl-

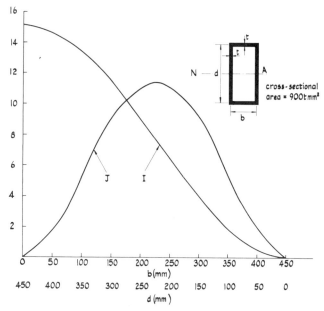

FIG. 10.1. Section design: I and J versus width and depth.

ing. The nominal thicknesses finally adopted were 4 mm for tension flange and webs and 6 mm for compression flange.

The beam dimensions and loading are shown in Fig. 10.2, which also indicates diaphragms under loading and support points with intermediate diaphragms at 500 mm centres, these diaphragms being incorporated as stiffeners to prevent premature buckling.[7]

10.3 SELECTION OF GRP MATERIAL

Theoretical analysis[10] suggests that the optimum fibre orientation to resist compression buckling is $\pm 45°$ to the direction of applied stress, whilst a 60° orientation is most effective in resisting buckling due to applied shearing forces. Experimental evidence[11] suggests, however, that multidirectional reinforcement at 0° and $\pm 45°$ gives the optimum resistance to compression buckling and unidirectional reinforcement at 45° is most efficient in resisting shear buckling. For resistance to shear buckling the 45° direction is chosen for the fibre, which coincides with the 'diagonal tension' direction.

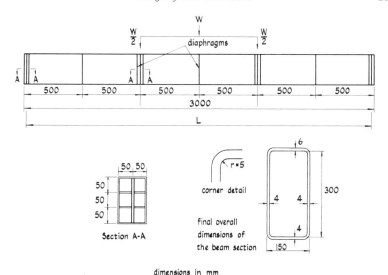

FIG. 10.2. General arrangement of box-beam.

In order to examine the part played by fibre orientation, six beams (a trial or pilot beam plus five others) were designed, manufactured and tested to destruction. To reduce the number of parameters involved the reinforcement in the webs and diaphragms of each beam were identical. In all cases the reinforcement was symmetrically layered with respect to the middle thickness of the flange, web or diaphragm.

Table 10.1 gives the details of the reinforcement used in each of the six beams, where the fibre orientation relates to the axes defined in Fig. 10.3. It should be noted that for web reinforcement the positive direction coincides with the 'diagonal tension' direction, i.e. it is measured clockwise at the left-hand end of the beam and anticlockwise at the right-hand end of the beam. Chopped strand mat was used in the corners of the beam cross-section to resist stress concentrations which would arise due to differing fibre orientations in flanges and webs. Bonding of diaphragms to webs and flanges was accomplished with CSM.

10.4 BEAM MANUFACTURE

The beams could not be cast in one operation due to the presence of diaphragms, so the following sequence of operations was adopted:

TABLE 10.1
DETAILS OF BEAM REINFORCEMENT

Beam	Web	Compression flange	Tension flange	Diaphragm
BP	4B 4(0/90)	4B 4(0/90)	4U 4(0)	4B 4(0/90)
B1	4U 0/+45/+45/0	6U 6(0)	UBBU 0/(0/90)(90/0)/0	4U 0/+45/+45/0
B2	4U 60/−45/−45/60	6U 6(0)	4U 4(0)	4U 60/−45/−45/60
B3	4U +45/−45/−45/+45	6U 6(0)	4U 4(0)	4U +45/−45/+45/−45
B4	6U 0/+45/−60/−60/+45/0	6U 6(0)	4U 4(0)	6U 0/+45/−60/−60/+45/0
B5	4B 4(0/90)	6B 6(0/90)	4B 4(0/90)	4B 4(0/90)

Abbreviations: B, bidirectional fibre; U, unidirectional fibre.

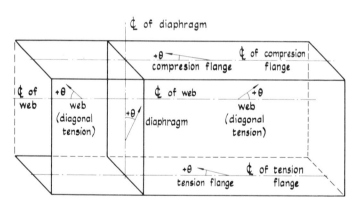

FIG. 10.3. Axes relative to fibre reinforcement orientation.

(a) Using a suitable mould, two layers of fibre were cast for the tension and compression flanges, and one layer of fibre for one web, thus forming a U-shaped cross-section.
(b) Diaphragms were cast to their finished thickness from a suitable mould.

(c) The diaphragms were placed in the U-shaped cross-section and bonded to it using CSM.
(d) One fibre layer of the remaining web was cast in a suitable mould, then placed on top of the diaphragms and bonded in with CSM.
(e) The remaining fibre layers, as detailed in Table 10.1, were cast.

Experimental methods were used to determine the elastic and strength properties of the GRP material used in the various elements of the beam cross-section. Material samples for this purpose were cut from the beams. These material properties, together with the average measured thickness of each beam element, are shown in Tables 10.2 to 10.7.

TABLE 10.2
STRUCTURAL PROPERTIES OF BEAM BP

Structural property	Web	Compression flange	Tension flange	Diaphragm
Average thickness (mm)	4·03	4·30	4·02	6
Modulus (kN/mm²)				
Longitudinal, E_L	15·60(T)	15·60(C)	15·80(T)	12·25(T)
Transverse, E_T	9·09(T)	13·21(C)	—	8·43(T)
Shear, G	4·02	4·02	—	3·89
At 45°	10·73(T)	—	—	9·17(T)
Poisson's ratio				
Longitudinal, v_L	0·13	0·17	—	0·13
Transverse, v_T	0·11	0·12	—	0·09
Transformed section				
Modulus (kN/mm²)	15·60	15·60	15·60	15·60
Equivalent thickness (mm)	4·03	4·30	4·07	4·71
Compressive strength (N/mm²)	—	125·00	—	—

Abbreviations: T, tensile; C, compressive.

10.5 TRANSFORMED SECTION AND CRITICAL BUCKLING STRESSES FOR DESIGN

The critical stresses for design purposes are those which occur in the compression flange, webs and diaphragms tending to produce local buckling, as failure by buckling occurs before the ultimate strength of the materials is realised.

As the various elements of the beam cross-section have differing modulus values, the beam must be analysed by the 'transformed cross-section' method. In this the actual thickness of an element is transformed

264 GRP in structural engineering

TABLE 10.3
STRUCTURAL PROPERTIES OF BEAM B1

Structural property	Web	Compression flange	Tension flange	Diaphragm
Average thickness (mm)	4·27	5·30	3·99	6·00
Modulus (kN/mm^2)				
Longitudinal E_L	16·80(T)	17·81(C)	17·70(T)	12·53(T)
Transverse, E_T	9·65(T)	4·95(C)	—	8·50(T)
Shear, G	5·34	4·02	—	4·25
at 45°	12·81(T)	—	—	10·31(T)
Poisson's ratio				
Longitudinal, v_L	0·32	0·27	—	0·28
Transverse, v_T	0·18	0·08	—	0·17
Transformed section				
Modulus (kN/mm^2)	17·81	17·81	17·81	17·81
Equivalent thickness (mm)	4·03	5·30	3·97	4·22
Compressive strength (N/mm^2)	—	198·41	—	—

Abbreviations: T, tensile; C, compressive.

TABLE 10.4
STRUCTURAL PROPERTIES OF BEAM B2

Structural property	Web	Compression flange	Tension flange	Diaphragm
Average thickness (mm)	4·12	5·23	4·09	6·00
Modulus (kN/mm^2)				
Longitudinal, E_L	8·10(T)	18·06(C)	18·62(T)	6·30(T)
Transverse, E_T	7·20(T)	5·54(C)	—	6·20(T)
Shear, G	9·13	4·02	—	2·10
At 45°	15·34(T)	—	—	6·28(T)
Poisson's ratio				
Longitudinal, v_L	0·45	0·25	—	0·43
Transverse, v_T	0·42	0·11	—	0·43
Transformed section				
Modulus (kN/mm^2)	18·06	18·06	18·06	18·06
Equivalent thickness (mm)	1·84	5·32	4·21	2·09
Compressive strength (N/mm^2)	—	191·20	—	—

Abbreviations: T, tensile; C, compressive.

into an equivalent thickness of material having the same modulus as the compression flange. Thus the thickness of the various elements are transformed in the following manner:

$$\text{Transformed web thickness} = \text{actual web thickness} \times \frac{E(\text{web})}{E(\text{compression flange})}$$

Design of GRP box-beams

TABLE 10.5
STRUCTURAL PROPERTIES OF BEAM B3

Structural property	Web	Compression flange	Tension flange	Diaphragm
Average thickness (mm)	4·10	5·26	4·14	6·00
Modulus (kN/mm^2)				
Longitudinal, E_L	8·20(T)	17·81(C)	17·80(T)	7·50(T)
Transverse, E_T	8·20(T)	5·93(C)	—	7·50(T)
Shear, G	5·77	4·02	—	5·15
at 45°	12·61(T)	—	—	11·21(T)
Poisson's ratio				
Longitudinal, v_L	0·41	0·26	—	0·39
Transverse, v_T	0·36	0·09	—	0·39
Transformed section				
Modulus (kN/mm^2)	17·81	17·81	17·81	17·81
Equivalent thickness (mm)	1·89	5·26	4·14	2·53
Compressive strength (N/mm^2)	—	201·30	—	—

Abbreviations: T, tensile; C, compressive.

with similar expressions for the tension flange and diaphragm. The transformed section, which is considered to be composed wholly of compression flange material, is thus theoretically equivalent to the actual cross-section. The theory of bending can be applied to the transformed section to calculate such quantities as second moment of area, neutral

TABLE 10.6
STRUCTURAL PROPERTIES OF BEAM B4

Structural property	Web	Compression flange	Tension flange	Diaphragm
Average thickness (mm)	5·10	5·28	3·98	6·00
Modulus (kN/mm^2)				
Longitudinal, E_L	15·10(T)	17·85(C)	17·85(T)	13·31 (T)
Transverse, E_T	7·34(T)	5·01(C)	—	6·40 (T)
Shear, G	6·49	4·02	—	4·62
At 45°	12·37(T)	—	—	9·64 (T)
Poisson's ratio				
Longitudinal, v_L	0·25	0·25	—	0·22
Transverse, v_T	0·12	0·09	—	0·13
Transformed section				
Modulus (kN/mm^2)	17·85	17·85	17·85	17·85
Equivalent thickness (mm)	4·31	5·28	3·98	4·47
Compressive strength (N/mm^2)	—	187·25	—	—

Abbreviations: T, tensile; C, compressive.

TABLE 10.7
STRUCTURAL PROPERTIES OF BEAM B5

Structural property	Web	Compression flange	Tension flange	Diaphragm
Average thickness (mm)	4·12	5·29	4·18	6
Modulus (kN/mm^2)				
Longitudinal, E_L	14·10(T)	13·45(C)	13·93(T)	11·94(T)
Transverse, E_T	13·98(T)	13·45(C)	—	11·90(T)
Shear, G	3·46	4·02	—	3·34
At 45°	9·68(T)	—	—	8·95(T)
Poisson's ratio				
Longitudinal, v_L	0·13	0·16	—	0·12
Transverse, v_T	0·13	0·16	—	0·12
Transformed section				
Modulus (kN/mm^2)	13·45	13·45	13·45	13·45
Equivalent thickness (mm)	4·32	5·29	4·33	5·33
Compressive strength (N/mm^2)	—	141·6	—	—

Abbreviations: T, tensile; C, compressive.

axis position, deflection and stress. It should be noted, however, that a calculated stress in the transformed section must be converted back to relate to the actual material used in the actual section, thus:

$$\frac{\text{Stress in web of}}{\text{actual section}} = \frac{\text{calculated stress in}}{\text{web of transformed section}} \times \frac{E(\text{web})}{E(\text{compression flange})}$$

TABLE 10.8
PROPERTIES OF TRANSFORMED SECTION

Beam	$10^{-6} \times$ Second moment of area I (mm^4)	Distance of neutral axis from tension flange (mm)	Distance of neutral axis from compression flange (mm)	$10^{-3} \times$ Section modulus Z_T (mm^3)	$10^{-3} \times$ Section modulus Z_C (mm^3)
BP	46·31	152	148	304·7	312·9
B1	49·14	158	142	311·0	346·1
B2	36·32	159	141	228·4	257·6
B3	40·10	163	137	246·0	292·7
B4	50·49	157	143	321·6	353·1
B5	51·85	155	145	334·5	357·6

The transformed web, tension flange and diaphragm thicknesses for the beams are shown in Tables 10.2 to 10.7 calculated on the above basis. Other geometrical section properties of the transformed cross-section are shown in Table 10.8.

Buckling stresses in the compression flange, web and diaphragm can be determined theoretically[12] as follows:

(a) *Compression flange*

$$\sigma_{cr} = \frac{4\pi^2}{b^2 t}(\sqrt{D_L D_T} + H)$$

where

$$D_L = \frac{E_L t^3}{12(1 - v_L v_T)}$$

$$D_T = \frac{E_T t^3}{12(1 - v_L v_T)}$$

$$H = \frac{1}{2}(v_L D_T + v_T D_L) + \frac{2 G t^3}{12(1 - v_T v_L)}$$

(b) *Web*

$$\tau_{cr} = \frac{4K \sqrt[4]{D_L D_T^3}}{b^2 t} \quad \text{for } \theta > 1$$

or

$$\tau_{cr} = \frac{4K \sqrt{D_T H}}{b^2 t} \quad \text{for } \theta < 1$$

where $\theta = \sqrt{D_L D_T}/H$ and the value of K depends on θ as shown in Table 10.9.

(c) *Diaphragms*

$$\sigma_{cr} = \frac{k \pi^2 \sqrt{D_L D_T}}{b^2 t}$$

where

$$k = \frac{8}{3}\left(3 + \frac{0 \cdot 88 H}{\sqrt{D_L D_T}}\right)$$

TABLE 10.9
VALUES OF θ AGAINST K IN WEB

θ	K	θ	K
0	18·6	3·0	17·6
0·2	18·9	5·0	16·6
0·5	19·9	10·0	15·9
1·0	22·2	20·0	15·5
2·0	18·8	40·0	15·3

10.6 BEAM STRESSES

Under a total applied load W kN, the various stresses in the beam may be calculated in the following manner:

(a) *Compression flange*

$$\text{Mid-span stress} = \sigma_c = \frac{M}{Z_c}$$

where

$$M = 1000 \cdot \frac{W}{2} = 500W \text{ kN mm}$$

and

Z_c is the compression section modulus of the transformed section and has the value given in Table 10.8.

Hence

$$\sigma_c = \frac{5 \times 10^5 \, W}{Z_c} \text{ N/mm}^2$$

(b) *Tension flange*
Similarly the mid-span stress is the tension flange is:

$$\sigma_T = \frac{5 \times 10^5 \, W}{Z_T} \text{ N/mm}^2$$

where Z_T is the tension section modulus of the transformed section and has the value given in Table 10.8.

(c) *Web*
The maximum shearing stress in the web may be calculated from:

$$\tau = \frac{V A \bar{y}}{I t}$$

where V = shearing force = $W/2$ kN, $A\bar{y}$ = first moment of area of transformed section above the neutral axis taken about the neutral axis, I = second moment of area of transformed section, t = transformed thickness of two webs.

Alternatively, as an approximation, the maximum shearing stress in the web may be taken to be equal to 1·5 times the average shearing stress in the web, thus:

$$\tau = \frac{3}{2} \cdot \frac{W}{2} \cdot \frac{1}{d.2t_w} \text{ kN/mm}^2$$

where t_w = actual thickness of one web and d = depth of beam = 300 mm. Hence:

$$\tau = \frac{3 \times 10^3 \, W}{8 d t_w} \text{ N/mm}^2$$

(d) *Diaphragms*
The most heavily loaded diaphragms, i.e. those under the loads or support reactions may be considered to support the entire load or reaction. Hence the compressive stress in the diaphragm would be:

$$\sigma_D = \frac{W}{2.bt_D} \text{ kN/mm}^2$$

where t_D = actual diaphragm thickness and b = breadth of beam = 150 mm. Hence:

$$\sigma_D = \frac{10^3 \, W}{2 b t_D} \text{ N/mm}^2$$

The above stresses were calculated for a value of W equal to the observed failure load of the beams and are shown in Table 10.10. Also shown in

TABLE 10.10
BEAM STRESS AT FAILURE

Beam	Failure load, W (kN)	Calculated stresses at failure load (N/mm²)				Theoretical buckling stresses (N/mm²)			Ultimate strength of flanges (N/mm²)
		Tension flange	Compression flange	Web	Diaphragm	Compression flange	Web	Diaphragm	
BP	42·5	69·7	67·9	13·2	23·8	67·9	56·0	138·0	125·0
B1	60·0	96·5	86·7	17·6	33·6	79·2	67·0	141·0	198·4
B2	50·0	109·5	97·0	15·2	28·0	84·5	41·0	82·0	191·0
B3	52·3	106·3	89·3	15·9	29·3	82·6	46·0	103·0	201·3
B4	65·0	101·1	92·0	15·9	36·4	79·8	76·0	126·0	187·3
B5	62·0	92·7	86·7	18·8	34·7	99·7	78·0	163·0	141·6

Table 10.10 are the theoretical buckling stresses calculated by the methods described in section 10.5 and the ultimate strength of the material of the flanges.

From Table 10.10 is clear that the design calculations indicate that failure of all six beams will occur by buckling of the compression flange as the calculated stresses are approximately equal to the theoretical buckling stresses in the compression flange. It is also evident from Table 10.10 that no other mode of failure is even remotely likely. The calculated stresses in the web and diaphragm are well below their respective theoretical buckling stresses. Yield or fracture of the flange material will not occur as the ultimate strength of the flange material is considerably in excess of the calculated flange stresses at failure.

10.7 EXPERIMENTAL BEHAVIOUR OF THE BEAMS

The experimental behaviour of the beams was as predicted by the design calculations summarised in Table 10.10. All the beams failed by buckling of the compression flange. Beams B1 to B4 showed buckling waves before the failure load was achieved and it was possible to increase the load further before total failure occurred; this is consistent with the values in Table 10.10 (i.e. calculated compression flange stresses > theoretical buckling stresses). By contrast, in beams BP and B5 formation of buckling waves and failure were coincident. This is again consistent with the values of Table 10.10 (i.e. calculated compression flange stresses $\leq$ theoretical buckling stresses).

The theoretical mid-span deflection of the beams may be calculated from:

$$\delta = \frac{W a(3L^2 - 4a^2)}{48 EI}$$

where W is the applied load, L is the span, a is the distance between load and support (1000 mm), I is the second moment of area of the transformed section and E is the modulus of elasticity of the transformed section. Experimental and theoretical deflections are compared in Table 10.11, from which it may be seen that there is good agreement between calculated and experimental deflections except for beam B2 where the experimental deflection was only two-thirds of that predicted.

TABLE 10.11
MID-SPAN DEFLECTIONS AT AN APPLIED LOAD $W = 1\,\text{kN}$

Beam	Theoretical deflection (mm)	Experimental deflection (mm)	Experimental / Theoretical
BP	0·66	0·65	0·98
B1	0·55	0·60	1·09
B2	0·73	0·50	0·68
B3	0·67	0·65	0·97
B4	0·53	0·45	0·85
B5	0·69	0·70	1·01

10.8 EFFECT OF GRP MATERIAL PROPERTIES ON BEAM PERFORMANCE

The choice of type of reinforcement affects the GRP material properties which in turn affect the beam performance in a number of ways. These are considered below.

10.8.1 Modulus of Elasticity

The highest longitudinal modulus of elasticity was achieved in the compression flange when this was reinforced with unidirectional reinforcement having a longitudinal orientation (see Tables 10.2 to 10.7). Such reinforcement gave a modulus value of about $18\,\text{kN/mm}^2$ compared with the lowest value of about $15\,\text{kN/mm}^2$ for bidirectional reinforcement. This higher value of longitudinal modulus is advantageous as it will result in reduced deflections. It will not, however, necessarily give an improved buckling resistance. From the theoretical expressions given earlier it will be seen that the critical buckling stress depends on $\sqrt{E_L E_T}$. Although unidirectional reinforcement gives the highest value of longitudinal modulus (E_L) it gives a low value of transverse modulus (E_T). Thus comparing beams B1 and B5 in this respect it is seen that the unidirectional reinforcement of beam B1 has a value of:

$$\sqrt{E_L E_T} = \sqrt{17\cdot81 \times 4\cdot95} = 9\cdot39\,\text{kN/mm}^2$$

whilst the bidirectional reinforcement of beam B5 has a value of:

$$\sqrt{E_L E_T} = \sqrt{13\cdot45 \times 13\cdot45} = 13\cdot45\,\text{kN/mm}^2$$

Although beam B5 has a lower longitudinal modulus than B1, because of its superior transverse modulus it has the better performance in resisting compression buckling (Table 10.10—99·7 N/mm² against 79·2 N/mm²).

10.8.2 Compressive Strength
The material compressive strength was highest for unidirectional reinforcement in line with the stress direction (approximately 200 N/mm²) and lowest for the bidirectional material (approximately 140 N/mm²).

10.8.3 I Value of Transformed Section
The I value of the transformed section affects both the beam stresses and deflections. The higher the I value the lower the stresses and deflections. Beam B2 exhibited the lowest I value ($36·32 \times 10^6$ mm⁴; Table 10.8) primarily because the GRP material used in the webs of this beam (0°, 45°, 45°, 0° orientation) reduced the actual web thickness of 4·12 mm to an equivalent web thickness of only 1·84 mm. Beam B3 with web reinforcement orientation ($+45°$, $-45°$, $+45°$, $-45°$) shows a similar reduction in equivalent web thickness and hence in I value. In contrast beams B1, B4 and B5 have equivalent web thicknesses approximately equal to their actual web thicknesses and thus correspondingly high I values (approximately 50×10^6 mm⁴; Table 10.8).

10.8.4 Prevention of Compression Buckling Failure
In order that the beams would reach the ultimate compressive strength of the GRP flange material before failing prematurely in compression buckling, either the longitudinal and transverse moduli of the flange material or the compression flange thickness would have to be increased.

Consider, for example, beam B1 which has a compression flange thickness of 5·3 mm, a buckling stress of 79·2 N/mm² and an ultimate compressive strength of 198·4 N/mm² (Table 10.10). If the compression flange thickness of this beam was to be increased to t_1 so that yielding rather than buckling failure would occur, then:

$$\sigma_{cr} \geqslant 198·4$$

$$79·2 \left(\frac{t_1}{5·3} \right)^2 \geqslant 198·4$$

$$t_1 \geqslant 8·5 \text{ mm}$$

10.9 BEHAVIOUR OF BEAMS UNDER LONG TERM LOADING

So far in this chapter the design and experimental behaviour of GRP beams under short term loading has been discussed. Consideration will now be given to their behaviour under long term loading.

The beam cross-section chosen for long term loading tests was that of beam B5. Beams of this cross-section were manufactured but with double the previous span, i.e. 6 m rather than 3 m. The span was increased so that an excessive amount of dead load would not be required to load the beams experimentally. Third span point loading was again used.

The short term failure load for beam B5 was 62·0 kN (Table 10.10), corresponding to a mid-span bending moment at failure of 31 000 kN mm. Recommended load factors for short term and long term loading are 2 and 4 respectively. Thus for long term loading the working load bending moment would be 7750 kN mm, which on a 6 m span with third span point loading gives a total applied load of 7·75 kN. In the event it was decided to apply a load somewhat higher than this to give a more severe test condition. A 10 kN load was applied, divided equally between third span points, using dead load so that the load remained constant as creep deflection of the beam occurred. This loading corresponds to a load factor of 3·1 on the failure load.

Three nominally identical beams were investigated under the following conditions:

(a) Beam R1 was subjected to a permanent imposed load of 10 kN for 16 000 hours (approximately two years); for the first month strain and deflection readings were taken at frequent intervals, thereafter at one-monthly intervals.
(b) Beam R2 was subjected to a load–unload cycle for three months and thereafter to a permanent load of 10 kN for the remainder of the two-year period. Strain and deflection readings were taken as in (a).
(c) Beam R3 was subjected to its own self-weight (approximately 0·5 kN) and readings taken as in (a).

The beams were housed in a lower basement where humidity and temperature variations were minimal. Measurements of deflection and strain in the compression and tension flanges were made at mid-span, whilst shearing strain in the web was measured at the supports. The initial values of these quantities, at time $t=0$, are shown in Table 10.12.

If the total deflection at some time t is δ, then the deflection due to creep is $(\delta - \delta_0)$ where δ_0 is the initial deflection at time $t=0$. Similarly

TABLE 10.12
INITIAL VALUES OF DEFLECTION AND STRAIN

Initial deformation or strain $t=0$	Beam R_1	Beam R_2	Beam R_3
Deflection, δ_0 (mm)	32·1	33·1	2·5
Tensile strain, ε_0	$1·2 \times 10^{-3}$	$1·4 \times 10^{-3}$	$2·0 \times 10^{-4}$
Compressive strain, ε_0	$1·4 \times 10^{-3}$	$1·4 \times 10^{-3}$	$1·7 \times 10^{-4}$
Shear strain, γ_0	$1·2 \times 10^{-3}$	$1·3 \times 10^{-3}$	$2·0 \times 10^{-4}$

the tensile or compressive creep strain is $(\varepsilon - \varepsilon_0)$ and the shear creep strain is $(\gamma - \gamma_0)$. Values of these creep deflections and strains plotted to a natural scale are shown in Figs. 10.4 to 10.7. From these figures the following observations can be made.

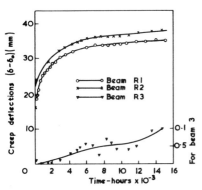

FIG. 10.4. Creep deflection at midspan vs. time.

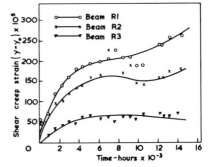

FIG. 10.5. Tensile creep strain vs. time.

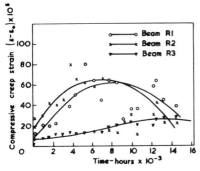

FIG. 10.6. Compressive creep strain vs. time.

FIG. 10.7. Shear creep strain at support vs. time.

(a) Initial deflections under approximately one-third ultimate load increased by 110% over a period of about two years. The majority (three-quarters) of this creep deflection occurred in the first 1000 h.
(b) Tensile strain in the tension flange at mid-span also increased by about 110% over a period of two years. About half of this tensile creep occured in the first 4000 h.
(c) Data for the compressive strain in the compression flange at mid-span were scattered, particularly after 8000 h. Compressive creep strains increased up to 8000 h and then decreased for the remainder of the two-year period. This is due to the modulus of elasticity of the compression flange material decreasing with time, resulting in the onset of compression buckling.
(d) Shearing strains over the two-year period increased by about 150%. About half of this increase occurred in the first 4000 h. The rate of shear creep declined towards the latter part of the two-year period due to the onset of buckling.

In order to derive expressions which can be used to predict creep effects, the natural plots of Figs. 10.4 to 10.7 were redrawn as log–log plots as shown in Figs. 10.8 to 10.11. From these plots three equations were derived to describe creep behaviour, namely:

$$\Delta = \Delta_0 + m\left(\frac{t}{t_0}\right)^n \tag{10.5}$$

$$\Delta = \Delta_0 + A \log_e\left(\frac{t}{t_0}\right) \tag{10.6}$$

$$\Delta = \Delta_0 + B \log_e\left(\frac{t}{t_0}\right) + C \tag{10.7}$$

where Δ represents the total deformation or strain at time t; Δ_0 represents the initial deformation or strain immediately after the load is applied; t_0 is a constant taken as unity and having the dimension of time; and m, n, A, B and C are constants which depend on stress, material and temperature.

The test data for creep properties for beam R1 (the beam which was continuously loaded) were compared with the above three equations for the best fit over the first 2000 h and then compared over the whole two-

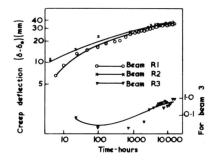

FIG. 10.8. Log–log plot of creep deflection vs. time.

FIG. 10.9. Log–log plot of tensile creep strain vs. time.

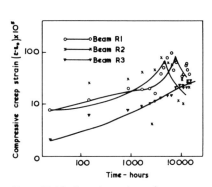

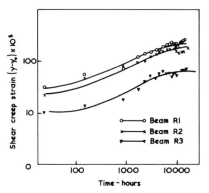

FIG. 10.10. Log–log plot of compressive creep strain vs. time.

FIG. 10.11. Log–log plot of shear creep strain at support vs. time.

year period of the test as shown in Figs. 10.12 to 10.15. Accordingly the values of the constants m, n, A, B and C were established. Comparing the experimental behaviour with the predictions of eqns. (10.5), (10.6) and (10.7) in this manner shows that:

(a) Equation (10.6) gives the best prediction of long term deflection creep behaviour with the constants in the equation having the following values:

$$\delta = \delta_0 + 4 \log_e \left(\frac{t}{t_0}\right) \qquad (10.8)$$

(b) Equation (10.5) gives the best prediction of long term creep in tensile strain with the constants in the equation having the following values:

$$\varepsilon = \varepsilon_0 + 10^{-5} \left(\frac{t}{t_0}\right)^{0.496} \qquad (10.9)$$

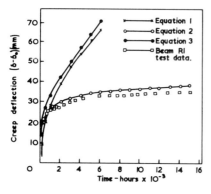

FIG. 10.12. Rectilinear plot of creep deflection at mid-span, beam R1, compared with the predictive equations.

FIG. 10.13. Rectilinear plot of tensile creep strain at mid-span, beam R1, compared with the predictive equations.

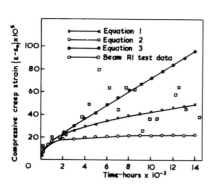
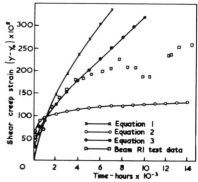
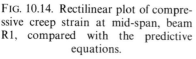

FIG. 10.14. Rectilinear plot of compressive creep strain at mid-span, beam R1, compared with the predictive equations.

FIG. 10.15. Rectilinear plot of shear creep strain at mid-span, beam R1, compared with the predictive equations.

From eqn. (10.9) it may be seen that $\log(\varepsilon - \varepsilon_0)$ is a linear function of $\log(t/t_0)$ with a slope of 0·496. This is confirmed by the straight line plot of experimental results shown in Fig. 10.9.

(c) None of the predictive equations shows good agreement with the experimentally observed compressive strain creep data. Up to 9000 h the best agreement is with eqn. (10.7) with constants in the equation having the following values:

$$\varepsilon = \varepsilon_0 + 47\cdot6 \times 10^{-5} \log_e \left(\frac{t}{t_0}\right) - 357\cdot0 \times 10^{-5}$$

(d) Up to a period of 5000 h eqn. (10.7) gives the best prediction for long term shearing strain creep with the constants in the equation having the following values:

$$\gamma = \gamma_0 + 148 \times 10^{-5} \log_e \left(\frac{t}{t_0}\right) - 1046 \times 10^{-5}$$

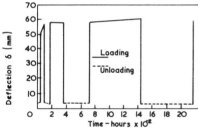

FIG. 10.16. Total deflection (elastic + creep), beam R2, during loading and unloading vs. time.

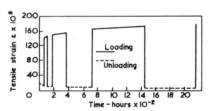

FIG. 10.17. Total tensile strain (elastic + creep), beam R2, during loading and unloading vs. time.

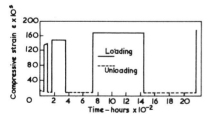

FIG. 10.18. Total compressive strain (elastic + creep), beam R2, during loading and unloading vs. time.

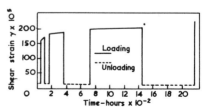

FIG. 10.19. Total shear strain (elastic + creep), beam R2, during loading and unloading vs. time.

The results for the beam (R2) subjected to alternating loads for a period of three months before being continuously loaded are shown in Figs. 10.16 to 10.19. From these figures it will be seen that the great majority of the creep strain or deflection was recovered during the unloading cycle.

REFERENCES

1. BENJAMIN, B. S., *Structural Design with Plastics*, 2nd edn., 1982, Van Nostrand Reinhold, New York.
2. HOLLAWAY, L., *Glass Reinforced Plastics in Construction*, 1978, Surrey University Press, Glasgow.
3. PIANO, R., Experiments and projects with industrialized structures in plastic materials, in *Int. Conf. on Space Structures, 1966*, 1967, Blackwell Scientific Publications, London.
4. BANKS, W. M. and RHODES, J., The buckling behaviour of reinforced plastic box sections, *The Reinforced Plastics Congress '80*, Brighton, Nov. 1980.
5. BANKS, W. M. and RHODES, J., The post buckling behaviour of composite box sections, in *Composite Structures*, ed. I. H. Marshall, 1981, Applied Science Publishers, London.
6. HOLDSWORTH, A. W. and OWEN, M. J., Macroscopic fracture mechanics of glass reinforced polyester resin laminates, *J. Composite Materials*, **8**, April 1974, 117.
7. SCHWARTZ, R. T. and ROSATO, D. V., Structural-sandwich construction, in *Composite Engineering Laminates*, ed. A. G. H. Dietz, 1969, MIT Press, Cambridge, Mass.
8. HOLMES, M. and RAHMAN, T. A., Creep behaviour of glass reinforced box beams, *Composites*, **11**, April 1980, 79.
9. RAHMAN, T. A., *Structural Behaviour of Glass Reinforced Plastics*, Ph.D. thesis, University of Aston in Birmingham, 1979.
10. ROTHWELL, A., Optimum fibre orientation in plastics and polymers, *Fibre Science and Technology*, **2**, 1969, 111.
11. AL-KHAYATT, Q. J., *The Structural Properties of Glass Fibre Reinforced Plastics*, Ph.D. thesis, University of Aston in Birmingham, 1974.
12. TIMOSHENKO, S. P. and GERE, J. M., *Theory of Elastic Stability*, 2nd edn., 1961, McGraw-Hill, New York.

CHAPTER 11

Design of a Stressed Skin Roof Structure

11.1 INTRODUCTION

The stressed skin structure is, like the box-section, a particularly suitable form for design in glass-reinforced plastics due to the fact that its stiffness is derived from its shape,[1–3] and a number of investigations into the behaviour of such forms have been undertaken.[4–6] In addition it has been shown in Chapter 9 that because of the specific strength and stiffness properties of GRP the material is particularly well suited to long span structures supporting relatively light loads.

This chapter will therefore be devoted to the design of a lightly loaded stressed skin roof, the structure shown in Fig. 11.1. Tests on a 1:6 scale model of the structure[7] confirmed that its behaviour was in agreement with design predictions.

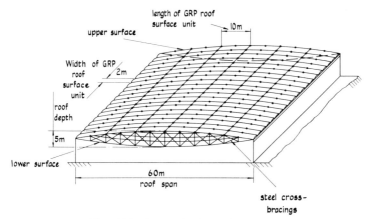

FIG. 11.1. Stressed skin roof structure.

282   *GRP in structural engineering*

The upper and lower parabolic surfaces are formed from a 10 m × 2 m GRP module (Figs. 11.2 and 11.3) which may be cast from a single mould. The module has double curvature which enhances its compression buckling resistance, and consists of a skin with stringer and ring

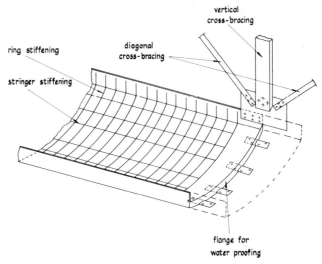

FIG. 11.2. View of lower surface unit joint region.

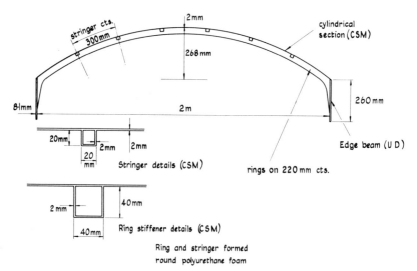

FIG. 11.3. Roof surface unit: cross-sectional details.

stiffeners in GRP with CSM reinforcement together with, for increased structural efficiency, edge beams reinforced with unidirectional (UD) fibres.

The upper and lower GRP surfaces are connected through a steel cross-bracing system, giving a variable depth structure to control the magnitude of the deflections. Two functions are served by the upper GRP surface, those of a stressed structural skin and of a cladding panel.

11.2 LOADING AND MATERIAL PROPERTIES

The structural loadings assumed were:

Self-weight 0·28 kN/m^2

Snow load 0·75 kN/m^2

Wind speed 33·6 m/s

These loadings are shown diagrammatically in Fig. 11.4 for the load combinations (wind + self-weight) and (snow load + self-weight). Bearing in mind that the structure is symmetrical about a horizontal axis, the critical load combination is (snow load + self-weight.)

Glass fibre reinforcement in the basic module was varied to take advantage of the particular properties required of the GRP material. The module edge beams were primarily required to resist axial force and these were reinforced with UD fibres with a fibre volume fraction of 54%. The remainder of the module, where local instability was a critical factor, was reinforced with CSM, fibre volume fraction of 16·5%. Resistance to local buckling was also improved by the inclusion of longitudinal stiffeners (stringers) and ring stiffeners. The former restrict the wavelength of buckling and hence increase the buckling load and the latter reduce the buckling sensitivity due to geometrical imperfections by constraining buckling to occur in an axisymmetric mode.

The short term strength and stiffness properties of these two GRP materials were taken to be:

CSM material   200 N/mm^2 compression

75 N/mm^2 tension

UD material 350 N/mm^2 compression

640 N/mm^2 tension

$E_{CSM} = 7 \times 10^3$ N/mm^2

$E_{UD} = 37 \times 10^3$ N/mm^2

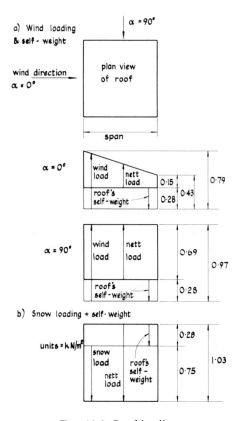

FIG. 11.4. Roof loading.

A structure life of 4×10^5 h (about 45 years) was assumed in determining the strength and stiffness properties at the end of the structure's useful life.

Experimental investigations[7,8] have indicated that the stresses which GRP material can sustain indefinitely are approximately equal to 40% of their short term ultimate strength properties. Consequently, the long term ultimate strengths of the two materials are assumed to be:

CSM material 80 N/mm² compression
              30 N/mm² tension
UD material   140 N/mm² compression
              260 N/mm² tension

It should be noted that the only stresses which are continuously applied to the structure are those due to self-weight. Therefore it is primarily the self-weight stresses which should be compared with the long term material strengths quoted above. Additional stresses, due to wind and snow loading, act for a matter of hours or weeks, rather than the 45 years over which it is assumed that the self-weight acts.

The effect of creep over the life of the structure may be expressed in terms of an apparent value of the modulus of elasticity[7] at the end of its life. This apparent modulus of elasticity may be calculated in the following manner.

If a stress σ is applied for a time t (h), then the strain at time t will be:

$$\varepsilon_t = \frac{\sigma_t n}{E_0} \qquad (11.1)$$

where E_0 is the apparent modulus of elasticity after 1 h of creep and n is a constant depending on the type of GRP material. It may be noticed that this equation and eqn. (8.15) are equivalent.

For the CSM and UD materials discussed here, eqn. (11.1) may be written:

$$\varepsilon_{t\,CSM} = \frac{\sigma_{CSM} t^{0.09}}{4200} \qquad (11.2)$$

Also,

$$\varepsilon_{t\,UD} = \frac{\sigma_{UD} t^{0.01}}{21\,000} \qquad (11.3)$$

$$\varepsilon_{t\,CSM} = \varepsilon_{t\,UD} \qquad (11.4)$$

It should be noted that eqns. (11.2), (11.3) and (11.4) apply for units of N/mm^2 only.

It is shown in section 11.3 that the axial force per unit width in the upper and lower GRP surfaces due to self-weight is 26·5 kN and that the cross-sectional areas per metre width of the CSM and UD material are 2500 mm^2 and 2100 mm^2 respectively. Hence equating the applied axial force to the sum of the axial forces taken by the two GRP components gives:

$$26\cdot5 \times 10^3 = 2500\,\sigma_{CSM} + 2100\,\sigma_{UD} \qquad (11\cdot5)$$

The solution of eqns. (11.2)–(11.5) for $t = 4 \times 10^5$ h gives the following results:

$$\varepsilon_{t\,CSM} = \varepsilon_{t\,UD} = 6 \cdot 31 \times 10^{-4}$$
$$\sigma_{CSM} = 0 \cdot 83 \text{ N/mm}^2$$
$$\sigma_{UD} = 11 \cdot 68 \text{ N/mm}^2$$

Hence the apparent moduli of elasticity of the two GRP materials after 4×10^5 h are:

$$\bar{E}_{CSM} = \frac{\sigma_{CSM}}{\varepsilon_{t\,CSM}} = 1 \cdot 31 \times 10^3 \text{ N/mm}^2$$

and

$$\bar{E}_{UD} = \frac{\sigma_{UD}}{\varepsilon_{t\,UD}} = 18 \cdot 48 \times 10^3 \text{ N/mm}^2$$

11.3 PRELIMINARY DESIGN

The approximate axial forces and stresses in the upper and lower GRP roof surfaces are first calculated so that suitable member cross-sectional areas can be established. Subsequently, a more precise analysis is carried out to establish the magnitude of secondary bending stresses.

From Fig. 11.4 the critical loadings are (self-weight) and (self-weight plus snow load). Both of these loadings are uniform and hence give rise to parabolic bending moment diagrams. As the depth of the roof structure also varies parabolically with span, the axial force in the GRP roof surface will be constant to a first approximation, and equal to:

$$\bar{N} = \frac{1 \cdot 05_w L^2}{8d} \text{ kN/m width}$$

where w is the load per unit area, L is the roof span, d is the roof depth at mid-span and an allowance of 5% is included for shearing force. Thus the axial force due to self-weight only is:

$$N_1 = \frac{1 \cdot 05 \times 0 \cdot 28 \times 60 \times 60}{8 \times 5} = 26 \cdot 5 \text{ kN/m width}$$

and that due to self-weight plus snow load is:

$$N_2 = \frac{1 \cdot 05 \times 1 \cdot 03 \times 60 \times 60}{8 \times 5} = 98 \cdot 0 \text{ kN/m width}$$

From Fig. 11.3 the relevant cross-sectional areas may be calculated as:

$$A_{CSM} = 2500 \text{ mm}^2/\text{m width}$$

$$A_{UD} = 2100 \text{ mm}^2/\text{m width}$$

Other required cross-sectional properties are best calculated by transforming the actual cross-section into an 'equivalent' all CSM cross-section. This transformation is carried out for the structure at the end of its life, when the apparent moduli of elasticity, which have already been established, are:

$$\bar{E}_{CSM} = 1.31 \times 10^3 \text{ N/mm}^2$$

$$\bar{E}_{UD} = 18.48 \times 10^3 \text{ N/mm}^2$$

The cross-sectional area of UD material expressed as a transformed area (A_{UD}^1) of CSM material is:

$$A_{UD}^1 = A_{UD} \frac{\bar{E}_{UD}}{\bar{E}_{CSM}}$$

$$= 2100 \times \frac{18.48}{1.31}$$

$$= 29\,600 \text{ mm}^2/\text{m width}$$

Thus the total transformed area of CSM material (A_{CSM}^1) is:

$$A_{CSM}^1 = A_{UD}^1 + A_{CSM}$$

$$= 29\,600 + 2500$$

$$= 32\,100 \text{ mm}^2/\text{m width}$$

The second moment of area of the transformed CSM section is:

$$I_{CSM} = 3.2 \times 10^8 \text{ mm}^4/\text{m width}$$

The depth to the neutral axis is:

$$n = 378 \text{ mm}$$

Therefore, due to self-weight only, the axial stresses in the upper and lower GRP surfaces are:

$$\sigma_{CSM} = \pm \frac{26.5 \times 10^3}{32\,100} = \pm 0.83 \text{ N/mm}^2$$

and

$$\sigma_{UD} \pm \frac{\sigma_{CSM} \bar{E}_{UD}}{\bar{E}_{CSM}}$$

$$= \pm \frac{0{\cdot}83 \times 18{\cdot}48}{1{\cdot}31}$$

$$= \pm 11{\cdot}68 \text{ N/mm}^2$$

These stress values are low compared with the stress levels which these materials can sustain indefinitely, i.e. $+30 \text{ N/mm}^2$ (tension) and -80 N/mm^2 (compression) for the CSM material and $+260 \text{ N/mm}^2$ and -140 N/mm^2 for the UD material.

Considering the imposition of a snow load in addition to self-weight, at the end of the life of the structure:

$$\sigma_{CSM} = \pm \frac{98{\cdot}0 \times 10^3}{32\,100} = \pm 3{\cdot}05 \text{ N/mm}^2$$

and

$$\sigma_{UD} = \pm \frac{3{\cdot}05 \times 18{\cdot}48}{1{\cdot}31} = \pm 43{\cdot}07 \text{ N/mm}^2$$

Compared with the stress values that these materials will sustain indefinitely, the above values show factors of safety of 9·8 and 3·2 for the CSM and UD materials respectively. These are considered adequate at the end of life of the structure, bearing in mind that the snow load is of a relatively short duration.

The above preliminary design calculations were supplemented by a more rigorous examination by computer analysis,[7] which was also used to examine shearing stresses and local buckling effects and is described in the next section.

11.4 COMPUTER ANALYSIS

The computer analysis indicated that the most highly stressed GRP module was adjacent to the support and was subjected to an axial force of 101·5 kN (preliminary design gave 98 kN) and a bending moment of 3·12 kN m (assumed zero in the preliminary design). The bending moments in other modules were only about 10% of this value. No doubt the

Design of a stressed skin roof structure   289

high bending stress in this module occurred because of the absence of a shear web in the panel adjacent to the support (see Fig. 11.1). The revised axial stresses for this module are:

$$\sigma_{CSM} = \pm 3{\cdot}16 \text{ N/mm}^2$$

$$\sigma_{UD} = \pm 44{\cdot}61 \text{ N/mm}^2$$

To these stresses must be added the bending stresses which are:

$$\sigma_{CSM} = \frac{M}{Z} = \frac{3{\cdot}12 \times 10^6 \times 378}{3{\cdot}2 \times 10^8}$$

$$= \pm 3{\cdot}68 \text{ N/mm}^2$$

$$\sigma_{UD} = \frac{M}{Z} \times \frac{\bar{E}_{UD}}{\bar{E}_{CSM}} = \frac{3{\cdot}12 \times 10^6 \times 150}{3{\cdot}2 \times 10^8} \times \frac{18{\cdot}48}{1{\cdot}31}$$

$$= \pm 20{\cdot}63 \text{ N/mm}^2$$

Thus the maximum combined stress in the CSM material is $\pm 6{\cdot}84 \text{ N/mm}^2$ and in the UD material is $\pm 65{\cdot}24 \text{ N/mm}^2$. Comparing these values with the stress that each of these materials will sustain indefinitely gives notional factors of safety of 4·4 and 2·2 for CSM and UD materials respectively. The true factors of safety are higher as the snow load is not acting 'indefinitely' and the safety factor at end of life is considered satisfactory.

The factors of safety inherent in these calculated stress values may be considered from another point of view. Given the short term strength properties of the materials, factors of safety of 2 and 4 may be used for short and long term loading respectively. Thus the safe working stresses under long term loading would be:

$$\text{CSM material } \frac{200}{4} = 50 \text{ N/mm}^2 \text{ compression}$$

$$\frac{75}{4} = 18{\cdot}75 \text{ N/mm}^2 \text{ tension}$$

$$\text{UD material } \frac{350}{4} = 87{\cdot}5 \text{ N/mm}^2 \text{ compression}$$

$$\frac{640}{4} = 160{\cdot}0 \text{ N/mm}^2 \text{ tension}$$

On this basis the stress values of $\pm 6.84 \, \text{N/mm}^2$ and $\pm 65.24 \, \text{N/mm}^2$ for the CSM and UD materials respectively are acceptable.

The computer analysis indicated an end of life maximum shearing stress of $0.065 \, \text{N/mm}^2$, which is insignificant. Mid-span deflection at end of life due to self-weight and snow load was 560 mm, i.e. about one hundredth of the span.

Euler and local buckling in the upper GRP surface must also be investigated. The Euler buckling load is:

$$P_E = \frac{\pi^2 EI}{L^2}$$

Thus for the transformed CSM cross-section at end of design life:

$$P_E = \frac{\pi^2 \times 1.31 \times 10^3 \times 3.20 \times 10^8}{(5000)^2} \, \text{N/m width}$$

$$= 1.65 \times 10^5 \, \text{N/m width}$$

$$= 165 \, \text{kN/m width}$$

giving a factor of safety of 1·63 which is adequate at end of design life.

The critical local buckling stress may be calculated[9] in the following manner for a symmetrically stiffened cylindrical surface:

$$N = \frac{2(1-v)^{1/2}}{r}(B_2 D_1)^{1/2}$$

where N is the buckling load/unit width, r is the cylinder radius and v is Poisson's ratio; also,

$$B_2 = E\left(t_s + \frac{A_r}{L_r}\right)(1-v^2)^{-1}$$

$$D_1 = EI_{st}(1-v^2)^{-1}$$

provided that $\quad \dfrac{B_1 D_2}{B_2 D_1} > 1.8 \text{ and } \dfrac{r}{h_r} < 100$

in which

$$B_1 = E\left(t_s + \frac{A_{st}}{L_{st}}\right)(1-v^2)^{-1}$$

$$D_2 = EI_r(1-v^2)^{-1}$$

and t_s is the skin thickness, A_r is the ring area, A_{st} is the stringer area, I_r is the ring second moment of area/unit width, I_{st} is the stringer second moment of area/unit width, L_{st} is the stringer pitch, L_r is the ring stiffener pitch and h_r is the ring depth.

The proposed design meets the above conditions and gives a value of:

$$N = -31 \text{ kN/m width}$$

i.e. a buckling stress of:

$$\sigma = -\frac{31 \times 10^3}{2500} = -12\cdot4 \text{ N/mm}^2$$

The maximum compressive stress (axial plus bending stress) in the most highly loaded GRP module at the end of structure life is $-6\cdot84 \text{ N/mm}^2$; hence the factor of safety against local buckling is 1·81, which is satisfactory.

Load tests carried out on a 1:6 scale model of the above design showed that the general experimental behaviour was consistent with the design calculations and that the load bearing capacity and stiffness of the structure predicted by the design calculations were confirmed by the experiment.

REFERENCES

1. BENJAMIN, B. S., *Structural Design with Plastics*, 2nd edn., 1982, Van Nostrand Reinhold, New York.
2. HOLLAWAY, L., *Glass Reinforced Plastics in Construction*, 1978, Surrey University Press, Glasgow.
3. PIANO, R., Experiments and projects with industrialized structures in plastic materials, in *Int. Conf. on Space Structures, 1966*, 1967, Blackwell Scientific Publications, London.
4. GILKIE, R. C., A comparison between the theoretical and experimental analysis of a stressed skin system of construction in plastics and aluminium, in *Int. Conf. on Space Structures, 1966*, 1967, Blackwell Scientific Publications, London.
5. GILKIE, R. C., *Pyramids in Lightweight Roof Systems*, Ph.D. thesis, University of Surrey, 1967.
6. ROBAK, D., The structural use of plastic pyramids in double-layer space grids, in *Int. Conf. on Space Structures, 1966*, 1967, Blackwell Scientific Publications, London.
7. MOLYNEUX, K. W., *An Investigation into the Feasibility of the Structural Use of Glass Reinforced Plastics in Long Span Lightly Loaded Structures*, Ph.D. thesis, University of Aston in Birmingham, 1976.

8. BOLLER, K. H., The effect of long-term loading on glass-reinforced plastic laminates, *Proc. 14th Technical and Management Conf., Reinforced Plastics Division SPI*, Feb. 1959.
9. LAKSHMIKANTHAM, C. and GERARD, G., Minimum weight design of stiffened cylinders, *Th Aeronautical Quarterly*, Feb. 1970, 49.

Index

Accelerators, 20
Addition polymerisation, 12, 16
Additives, 43
Adhesive joints, 243–4
Aluminium and aluminium alloys, 8
Amorphous solids, 24
Anisotropic effects, 233–4
Anisotropic lamina, 87
Anisotropic materials, 42
Autoclave method, 37
Axial forces, 153
Axial stresses, 289

Bakelite, 12
Bending moment, 138, 140, 156, 187, 256, 288
Bessemer process, 7
Bidirectionally continuously reinforced composites, 224
Box-beams, 258–80
  compression flange, 267, 268
  critical stresses, 263
  cross-section, 259
  diaphragms, 267, 269
  effect of material properties on performance, 272–3
  experimental behaviour, 271
  general arrangement, 261
  I value of transformed section, 273
  loading, span and cross-sectional shape, 258–60

Box-beams—*contd.*
  long term loading, 274–80
  manufacture of, 261–3
  reinforcement, 261
  selection of material, 260–1
  short term design load, 258
  stiffening, 260
  stress
    analysis, 268–71
    failure, at, 270
  structural behaviour, 258
  structural design, 258
  structural properties, 263, 264
  tension flange, 268
  transformed section
    analysis, 263
    properties, 266
  web, 267, 269
Buckling
  failure, 273
  modes, 128
  stresses, 128, 248, 249, 267, 271, 290, 291

Cast iron, 6–7
Cement, 9
Cementing materials, 9
Chopped strand mat reinforced composites, 28, 222, 223, 224, 227, 228, 233, 235
Closed mould processes, 38–41

294　　　　　　　　　　　　　　　　　Index

Cold press mouldings, 40
Combination joints, 247, 248
Compliance matrix, 52
Composite laminae
　assumptions and idealisations made in theoretical treatment, 48–9
　macromechanical properties of, 47–104
　micromechanical analysis of, 105–33
　strength characteristics of, 87–103
　stress–strain relationships, 49
Composite materials, 6, 8–13
　history, 9
Compressive strength, 5, 10, 178, 182, 192–5, 207–9, 234–7, 273
Compressive stresses, 94, 291
Compressive tests, 173, 174
Concrete, 10
Condensation polymerisation, 12
Condensation reaction, 16, 19
Contact moulding, 33–5
Continuous laminating, 41
Continuous undirectional fibre lamina, 107–20, 125–30
Continuously reinforced laminates, 171–209
Corrosion properties, 7, 8
Creep
　behaviour, 23, 276
　effects, 48–9, 285
　properties, 4, 5, 13, 215, 217, 219, 220, 240, 276
　resistance, 223
　strain, 275
Critical aspect ratio, 122
Critical length, 122
Cross-linking, 18, 19
Curing process, 19–20

D glass, 25
Dashpot, 214, 216, 218, 220
Deformation
　modes, 128
　resistance, 3
Density requirements, 4
Directionally reinforced composites, 223

Discontinuous fibre lamina, 120–4, 130–2
Discontinuously reinforced laminates, 209–12
Distortion strain energy theory, 94–7
Distortion stress systems, 95
Double bonds, 14, 19
Ductile materials, 95, 97

E glass, 25, 27, 233
Elastic behaviour, 4
Elastic constants, 67
Environmental conditions, 23, 43
Equilibrium equation, 72
Ethylene, 11
　molecule of, 14
Extensional mode, 177

Fail safe materials, 4
Failure
　characteristics, 87
　criteria, 102, 103, 166
　load, 127
　mechanisms, 124–5
Fibre reinforcement materials, 23
Filament winding, 37–8
Fire penetration test, 242
Fire performance, 241–2
Flexural modulus, 238
Flexural strength, 237–8
Flexural tests, 173, 179–81
Force–deformation relationship, 4
Force–strain relationship, 198
Four-layered cross-ply laminate, 182–95
Fumaric acid, 17

Gel coat, 34
Glass
　components of, 24
　properties of, 25
　structure and types, 24–5
　two-dimensional structural representations, 25

Index

Glass fibre
 production process, 26
 properties, 27
 reinforced laminates, 134–70
 reinforcement, 23–30
  forms of, 28–30, 233
GRP
 applications, 13
 components of, 14–44
 material characteristics, 41–3
 methods of manufacture, 32–41
 philosophy of, 42

Hand lay-up, 33
Homogeneous materials, 6–8
Hot press moulding techniques, 39

Ignition tests, 242
Interlaminar stress effects, 168–9
Isotropic beams, 136–9
Isotropic laminae, 49–61
 strength hypotheses for, 89–97
Isotropic materials, 42, 55
Isotropic plate, 145–8

Joints, 243–7
 adhesive, 243–4
 combination, 247, 248
 mechanical, 244–7

Laminated beams, 136–43
 general equations for, 143
Laminated plates, 143–64
 general equations for, 161–4
Linear elastic material, 4
Linear viscoelastic models, 216
Linear viscoelasticity, 214–21
Linear viscosity coefficient, 215
Load–deformation characteristics, 43, 166, 184, 187, 190
Long term loading effects, 274–80
Long term strength and stiffness, 238–41

Macromechanical properties of composite laminae, 47–104
Maleic acid, 17
Manufacturing methods, 32–41
Matched die processes, 38–41
Material matrix, 52
Maximum principal strain theory, 90–1
Maximum principal stress theory, 90
Maximum shearing stress theory, 91–2
Maximum stress theory, 98–100
Maxwell model, 218–21, 223
Maxwell's reciprocal theorem, 53, 65
Mechanical joints, 244–7
Micromechanical analysis
 assumptions and limitations in, 106
 composite laminae, of, 105–33
Modulus of elasticity. *See* Young's modulus
Modulus of rigidity, 54, 63, 66, 67, 88, 105, 117, 175, 200, 211
Mohr's stress circle, 91
Moisture effects, 43
Moment–strain relationship, 198
Monobasic acid, 16
Monohydric alcohol, 16
Monomer addition, 19
Multidirectionally continuously reinforced laminates, 182–209
Multidirectionally reinforced composites, 171

Normal strains, 51, 144
Normal stress distribution, 147
Normal stresses, 51, 55, 58, 63–6, 102

Open mould process, 32–8
Orthotropic laminae, 61–87
 reinforcement forms, 63
 strength hypotheses for, 98–103
Orthotropic materials, 42, 67–70
Orthotropic plate, 149–52

Plastics, 11–13
Poisson's ratio, 21, 54, 56–8, 60, 63, 64, 66–70, 88, 89, 105, 107, 110, 113, 120, 152, 156, 174, 183, 185, 199, 200, 211
Polybasic acid, 16
Polyester
  forms of, 16
  molecule of, 16
  unsaturated, 17
Polyester matrix, 14–23
Polyester resins, 12, 14
  basic chemistry of, 14–18
  cured unsaturated, 20
  production of, 18–20
  unsaturated, 18–20
Polyethylene, 11
  formation of, 15
Polyhydric alcohol, 16
Polymers, 11
Polythene. See Polyethylene
Poly(vinyl chloride), 11, 12
Portland cement, 10
Preform, 39
Premix, 39
Pressure bag techniques, 35–7
Pultrusion process, 40–1

Quartz, 24

Relaxation, 215, 217, 219
Rovings, 30, 233
Rule of mixtures equation, 154

S glass, 25
Safety factor, 289
Second moment of area, 251
Self-weight effects, 249–51
Shear failure, 89, 90, 92, 94
Shear mode, 130, 177, 178
Shear modulus, 235, 238
Shearing displacement, 118
Shearing forces, 148
Shearing strain, 53–6, 101, 118, 144, 201
Shearing strength, 5, 181, 238

Shearing stress distribution, 147
Shearing stress–shearing strain relationship, 66, 70
Shearing stresses, 53, 60, 62, 66, 89, 92, 117, 122, 148, 169, 269
Shell roof, 5
Short term design load, 258
Short term flexural strength and stiffness, 237–8
Short term shearing strength and stiffness, 238
Short term tensile and compressive strength and stiffness, 234–7
Shrinkage, 20
Siemens process, 7
Silica, 24
Silicon dioxide, 24
Specific gravity, 4, 20, 27
Spray-up technique, 35
Spring and dashpot, 214, 216, 218, 220
Steel properties and applications, 7
Stiffness
  characteristics, 48, 49, 87, 107, 135–64, 183, 196–201, 209–12
  long term, 238–41
  short term, 234, 237–8
  time and temperature dependence, 221–6
  properties, 5, 13, 20–1, 24, 27, 165, 174, 179
  requirements, 3
Strain
  distribution, 121
  energy, 92–7, 101
  relationships, 75–87
  transformation equations, 83, 84
Strength
  characteristics, 42, 49, 124–32, 164–7, 174, 181, 187–95, 201–12
  composite laminae, of, 87–103
  determination of, 88
  long term, 238–41
  short term, 234
  time and temperature dependence, 226–8
  hypotheses
    isotropic laminae, for, 89–97
    orthotropic laminae, for, 98–103

Strength—*contd.*
  properties, 20–1, 27
  requirements, 3
Stress
  distribution, 123, 138
  effects, 23
  relationships, 72
  systems, 95
Stress–strain characteristics, 236
Stress–strain equations, 100
Stress–strain ratio, 21
Stress–strain relationships, 43, 49, 70, 71, 81–7, 90, 101, 105, 107–20, 120–4, 145, 214, 234
Stressed skin roof structure, 281–92
  computer analysis, 288–91
  cross-sectional details, 282
  loading and material properties, 283–6
  preliminary design, 286–8
Structural design, 233–57
Structural joints, 243
Structural materials
  historical background, 3–13
  nature of, 6–13
  requirements of, 3–5
  structural form, effects on, 5
Structural properties, comparison with other structural materials, 248–53
Styrene, 19
Super-polymers, 11

Temperature dependent characteristics, 21, 213–29, 241
  stiffness effects, 221–6
  strength effects, 226–8
Tensile creep tests, 221–4
Tensile failure, 89, 90
Tensile strain, 207
Tensile strength, 10, 176, 187–92, 201–7, 234–7
Tensile strength anisotropy, 234
Tensile stresses, 94
Tensile tests, 173, 174
Thermal properties, 241
Thermoplastics, 12, 16, 213

Thermosetting plastics, 12, 18, 213
Three-dimensional analysis, 56–61, 67
Three-layered symmetric laminate, 152–7
Three-layered symmetrically laminated beam, 139–41
Timber, 9
Time dependent characteristics, 213–29
  stiffness effects, 221–6
  strength effects, 226–8
Torque, 148, 156
Torsional second moment of area, 259
Total strain energy theory, 92–4
Transformed sections, 253–6, 263, 266
Transverse strain, 115
Tsai–Hill energy theory, 101–3
Tsai–Hill failure criterion, 103, 167, 189, 193, 204, 205, 207, 208
Two-layered 45° angle-ply laminate, 196–209
Two-layered non-symmetric laminate, 157–61
Two-layered non-symmetrically laminated beam, 141–2

Ultimate compressive strain, 195
Ultimate compressive strength, 127–30, 195, 211
Ultimate tensile strain, 192
Ultimate tensile strength, 125–2, 176, 192, 206, 211
Uniaxial compressive test, 91
Uniaxial tensile test, 89, 92, 102
Unidirectionally continuously reinforced composites, 222, 223, 226, 227, 228
Unidirectionally continuously reinforced laminates, 174–82
Unidirectionally reinforced composites, 171

Vacuum bag technique, 35
van der Waals forces, 15–16
Vinyls, 12
Viscoelasticity, 214
Voigt model, 215–18, 220–1, 223

Volume stress systems, 95
von Mises criterion, 95

Water absorption effects, 43
Wood, 9
Woven fabrics, 30, 233
Woven rovings, 30
Wrought iron, 6–7

Young's modulus, 21, 54, 58, 105
   composite laminae, 88
   continuous unidirectional fibre laminae, 107
   determination of, 112–16
   discontinuously reinforced laminates, 211
   E glass, 28

Young's modulus—*contd.*
   flexural tests, 179
   isotropic laminae, 56, 89, 124
   long term, 239
   longitudinal, 123
   multidirectionally continuously reinforced laminates, 183, 185
   orthotropic laminae, 63, 64, 66, 67
   three-layered symmetric laminate, 156
   transverse, 120
   two-layered 45° angle-ply laminate, 199, 200
   two-layered non-symmetrically laminated beam, 141
   unidirectionally continuously reinforced laminates, 174, 177
   variation
      between layers, 139
      with temperature, 225

DEPT. C.W. Eng.
O.N. 9055.
PRICE £44-20
ACCN. No. RB 4914